Thin Liquid Films

Theoretical and Mathematical Physics

The series founded in 1975 and formerly (until 2005) entitled *Texts and Monographs in Physics* (TMP) publishes high-level monographs in theoretical and mathematical physics. The change of title to *Theoretical and Mathematical Physics* (TMP) signals that the series is a suitable publication platform for both the mathematical and the theoretical physicist. The wider scope of the series is reflected by the composition of the editorial board, comprising both physicists and mathematicians.

The books, written in a didactic style and containing a certain amount of elementary background material, bridge the gap between advanced textbooks and research monographs. They can thus serve as basis for advanced studies, not only for lectures and seminars at graduate level, but also for scientists entering a field of research.

For further volumes:
www.springer.com/series/720

Ralf Blossey

Thin Liquid Films

Dewetting and Polymer Flow

 Springer

Ralf Blossey
CNRS USR 3078
Institut de Recherche Interdisciplinaire
Villeneuve d'Ascq Cedex
France

ISSN 1864-5879   ISSN 1864-5887 (electronic)
Theoretical and Mathematical Physics
ISBN 978-94-007-4454-7   ISBN 978-94-007-4455-4 (eBook)
DOI 10.1007/978-94-007-4455-4
Springer Dordrecht Heidelberg New York London

Library of Congress Control Number: 2012938859

Printed on acid-free paper

Springer is part of Springer Science+Business Media (www.springer.com)

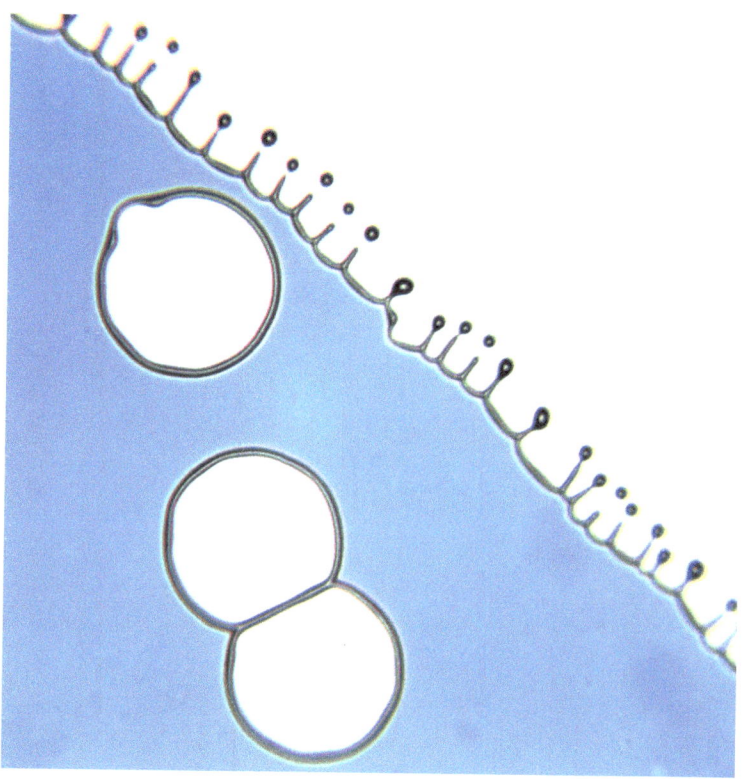

Two in one. This AFM picture displays two types of thin films instabilities at the same time: dewetting holes that open up in a thin polymer film and an instability of the film rim along its receding edge. This instability is similar to the classic Rayleigh instability of a liquid column. In the case of a thin film the polymer accumulates at the rim which upon becoming unstable produces a characteristic pattern which is reminiscent of a carpet fringe.

Courtesy: Karin Jacobs

How does it come that papers on wetting phenomena and other soft matter topics are often so confusing, while in condensed matter physics everything is so clear?

Vincent Senez, some time in 2010

Preface

This book is a treatise on the thermodynamic and dynamic properties of thin liquid films at solid surfaces and, in particular, their rupture instabilities. For the quantitative study of these phenomena, polymer thin films haven proven to be an invaluable experimental model system.

What *is* a *thin* liquid film? For the purpose of this book, thin films are (polymeric) liquids at surfaces whose properties are controlled by interfacial forces—capillary and intermolecular, like van der Waals forces. Gravity does not play a role for them. Some researchers prefer to call such films *ultrathin*.

What is it that makes thin film instabilities special and interesting, warranting a whole book? There are several answers to this. Firstly, thin polymeric films have an important range of applications, and with the increase in the number of technologies available to produce and to study them, this range is likely to expand. An understanding of their instabilities is therefore of practical relevance for the design of such films.

Secondly, thin liquid films are an interdisciplinary research topic. Interdisciplinary research is surely not an end to itself, but in this case it leads to a fairly heterogeneous community of theoretical and experimental physicists, engineers, physical chemists, mathematicians and others working on the topic. It justifies attempting to write a text which aims at a coherent, theoretical presentation of the field which researchers across their specialised communities might be interested in. It is in some sense a response to V. Senez' question: in solid state physics the community has much more converged to a common conceptual understanding, since people from a common scientific background work in it. But there is more: the wetting or soft matter field is dominated by an enormous diversity of phenomena and mostly experimental work (and seemingly simple theoretical explanations), apart from the theory of *wetting phase transitions*, which has a rigorous grounding in statistical physics. Thin liquid films are an interesting laboratory for a theorist to confront a well-established theory, hydrodynamics, with its limits. Liquids at surfaces take notice of the surface they are placed upon, and this is reflected in their dynamics. And the polymers, when confined to thin films, can imprint molecular properties on the film dynamics.

In the end, of course, we have only really learnt something about Nature when the theories have been confronted with reality. Here, again, lies a tremendous advantage in the case of thin polymeric films due to the modern experimental techniques with which they can be made and studied. This therefore is a field in which a highly fruitful exchange and collaboration is possible between experimentalists and theorists.

The material in the book is arranged in two Parts. Part I covers the basics of wetting and dewetting phenomena, and is of interest to researchers working in the field also outside of polymeric systems. It can be read as a brief introduction into the theory of wetting phase transitions. Part II delves exclusively into polymeric thin films, their mathematical description, and the confrontation with experiment. The exposition of this book is theoretical or mathematical in the sense that within each chapter and each section, calculations are presented at a great level of detail, but no proofs in any strict mathematical sense are given. For an experimental scientist, the book should serve as a reference and guide to what is the current consensus of the theoretical underpinnings of the field of thin film dynamics.

The field of wetting and dewetting owes a great debt to Pierre-Gilles de Gennes and his collaborators and students who were so influential for the field of Soft Matter Physics. Their work has produced deep insights which are, at the same time, presented in a mathematically 'light' and elegant fashion, often making use of scaling arguments. For the untrained, this approach is not always easy to follow. There is no point in trying to replace it with tedious technicalities, and this is *not* what is intended here. The present book attempts to bridge between the 'light' and the 'rigorous', always with the ambition to enhance insight and understanding—and to not let go the elegance of the theory.

This book owes a great deal to my collaborators and discussion partners over the years. I hope they all will find that it also reflects what I learnt from them.

Lille   Ralf Blossey
May 2012

Contents

List of Symbols[1]

Ca	capillary number
η	viscosity
σ, σ_{ij}	surface tension
ℓ_c	capillary length
ϱ	(liquid) density
g	gravitational acceleration
θ, θ_r	contact angle
S	spreading coefficient
$h(x)$	interface height/film thickness
$\mathcal{H}[h]$	(effective) interface Hamiltonian
$\Delta\mu$	chemical potential difference
τ	line tension
κ	(interfacial) mean curvature
Π_{vdW}	disjoining pressure
$V(h), \Phi(h)$	effective interface potential
$\hbar$	Planck's constant
c	speed of light
$\varepsilon_i(\omega)$	dielectric function
n_i	refractive index
A	Hamaker constant
ξ	decay length
Δ_r	radial Laplacian
E_c	excess free energy of a critical nucleus

---

[1]There is a particular difficulty in writing a book which covers, at the same time, the topics of thin liquid films *and* of hydrodynamics. The difficulty lies in notation. In thin films or wetting problems, σ or γ are used to denote surface tensions, while τ is the standard symbol of the line tension. But the very same symbols denote different kind of tensors in hydrodynamics or elasticity. This problem therefore cannot have a good solution. I hope that the one I chose for this book is, while certainly not perfect, at least a practical one. The main symbols used throughout the book are collected in this list for reference.

$H_c(R_c)$	height (radius) of a critical nucleus
κ_D^{-1}	Debye screening length
p	pressure
$\widehat{\sigma}$	stress tensor
$\widehat{\tau}$	extra-stress tensor
b	slip length
η	shear viscosity
Re^*, Re	(reduced) Reynolds number
$M(h)$	interface mobility
$M_\nu(A)$	Minkowski functionals
ω	vorticity
Z	enstrophy
$\widehat{F}$	deformation gradient tensor
$\widehat{\gamma}$	strain tensor
$\dot{\widehat{\gamma}}$	rate-of-strain tensor
ζ	friction coefficient
T_K	Kauzmann temperature

List of Figures

Part I
Dewetting

Chapter 1
Introduction

We start the book with a short look at two experiments in which surface forces and hydrodynamics play a role. In the first, polymers are not involved; it is done with oils. The second concerns a much more complex system, a *binary liquid mixture* of two polymers.

1.1 The Landau-Levich-Derjaguin Problem

The first experiment is an idealization of the '*dip-coating*' *process*: how to coat a surface by pulling it out from a bath of liquid that is supposed to leave a thin film on the surface. This is a classic problem of fluid mechanics which was first studied theoretically by Landau and Levich in 1942 and shortly afterwards by Derjaguin (1943).

The experimental setup is shown in Fig. 1.1(a) (which is in the center part of the figure). A flat plate is drawn out of a bath of a viscous liquid at a speed U such that a layer of liquid remains on the plate. In this problem, the question is how the thickness h of the coating layer is related to the speed U at which the plate is drawn out of the bath. The characteristic parameter here is the *capillary number*

$$Ca \equiv U\eta/\sigma \tag{1.1}$$

where η is the liquid viscosity and σ the surface tension of the liquid. Landau, Levich and Derjaguin solved the hydrodynamic problem in a lubrication approximation (an approach we will use extensively in Part II of the book) by imposing a smooth matching criterion on the plate moving with the film and a static meniscus at the entry into the bath. Such solutions are shown in Fig. 1.1(b). The size of the meniscus is controlled by the *capillary length*

$$\ell_c \equiv \sqrt{\sigma/\varrho g} \tag{1.2}$$

R. Blossey, *Thin Liquid Films*, Theoretical and Mathematical Physics,
DOI 10.1007/978-94-007-4455-4_1, © Springer Science+Business Media Dordrecht 2012

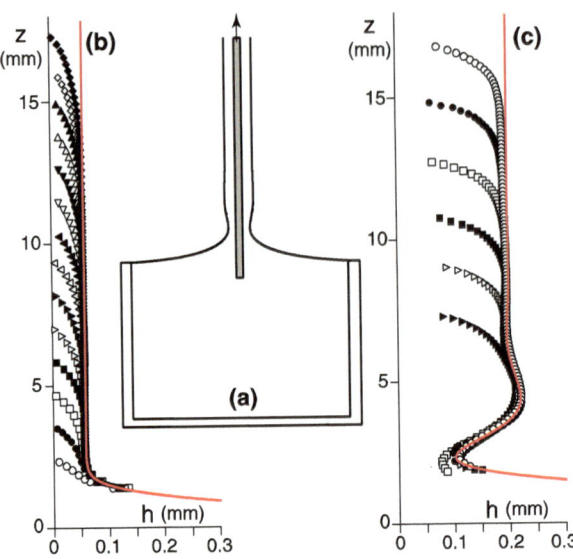

Fig. 1.1 *The Landau-Levich-Derjaguin dip-coating problem.* A flat plate is drawn out of a liquid bath (**a**). Two types of solutions are shown: (**b**) the LLD-solution with a smoothly matching meniscus, (**c**) another solution type in which the bath and the film on the plate are connected by a 'dimple' in the liquid film. Reprinted with permission from Snoeijer et al. (2008). Copyright by the American Physical Society

where ϱ is the liquid density and g the gravitational acceleration. In the limit of small capillary number Ca, i.e. for a 'not too fast' withdrawal of the plate, the sought relation for the film thickness turns out to be[1]

$$h_f \equiv \frac{h^{LLD}}{\ell_c} = 0.946 \, (Ca)^{2/3}. \tag{1.3}$$

This prototypical example of a dynamic wetting problem involving a 'thin' film has seen many variations, one of which is shown in Fig. 1.1(c). In fact, the matching criterion imposed by Landau, Levich and Derjaguin—the requirement that the meniscus be static—can be lifted, which then leads to another family of solutions. All solutions of the hydrodynamic problem can be characterized by the flux

$$q = h_f \left(Ca - \frac{h_f^2}{3} \right) \tag{1.4}$$

which has a parabolic shape, admitting for a given flux a thin film the LLD-solution and a thicker film. It is in the latter case that the meniscus displays a 'dimple'-like profile in the vicinity of the bath, as is clearly seen in Fig. 1.1(c).

---

[1] The power-law dependence on capillary number with an exponent of 2/3 was seen in experiments by Morey (1940), but indications to a nonlinear dependence were seen much earlier. A more detailed discussion of the history of the subject as well as a simple scaling argument to explain the exponent can be found in de Ryck and Quéré (1996).

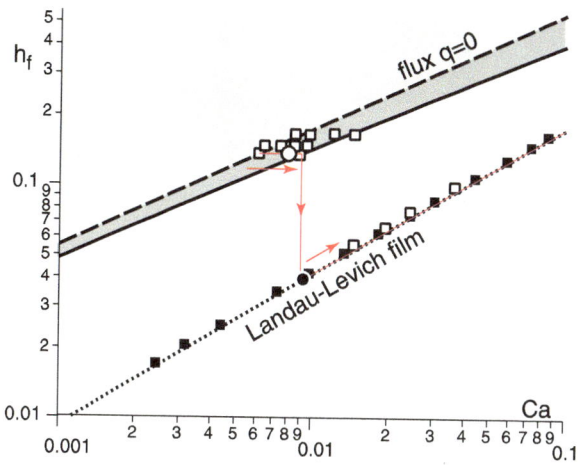

Fig. 1.2 A 'phase'-diagram of stationary solutions to the LLD-problem, locating both families (**b**) and (**c**). Solutions of type (**c**) exist in a gray-shaded area above the LLD-line. Experimental points are indicated in full (complete wetting) and open (partial wetting). The *red path* indicates the progression of the film thickness upon an increase in pull-out speed. The solutions jump from type (**c**) to LLD-type. Reprinted with permission from Snoeijer et al. (2008). Copyright by the American Physical Society

In the experiments which were compared with this theory, the dimple solutions were produced by changing the wetting properties of the surface: in the classic LLD-case, the liquid wets the plate 'completely' while in the modified version, it wets only 'partially'. These notions refer to a balance of surface tensions between the liquid, vapour and solid phases. They are a key property underlying the physics discussed in this book and will be explained in detail in Chap. 2.

Figure 1.2 summarizes the results, from both theory and experiment, in a kind of 'phase'-diagram in the parameter space of the thickness of the stationary film h_f and the capillary number Ca, using logarithmic scales. The thinner LLD-film follows a linear behaviour as given by equation (1.3). The thicker solutions lie in a window indicated as a grey-shaded area in the figure.

The excellent agreement between theory and experiment in this first example is not fortuitous: it should rather be taken as a key feature of the quality of the whole field in which a very close interaction between theory and experiment exists. In this book we will time and again encounter situations in which, for quite complex situations, such an excellent agreement has been achieved. But of course we will also encounter situations in which experiment and theory do not (yet) agree. There are, as always, two reasons for this. Either the theory is not yet up to the point, or the experiments are not yet sufficiently precise and controlled to allow for a result of a quality similar to the one shown in Figs. 1.1 and 1.2: an accord between theory and experiment without fitting parameters.

The following example has therefore been chosen to illustrate some aspects of the complexity inherent to the field; in this case, no such clear-cut accord between theory and experiment is so far available—simply because several factors compete.

1.2 Dewetting of a Liquid Film of a Binary Mixture

In the dip-coating experiment, the surface wetting properties of the plate help to select either the thin- or the thick-film solution, but they are not critical parameters for the appearance of the phenomenon itself. In the following example surface properties also matter, but the system is already vastly more complex in bulk.

In a paper published in *Science* in 1999, Yarushalmi-Rozen et al. presented results on a dewetting experiment performed on a partially miscible binary liquid mixture of two polymers, dOS, *deuterated oligomeric styrene*, and OEP, *oligomeric ethylene-propylene*, which have molecular weights of 580 and 2000 (see Appendix A for information about the properties of the various polymers appearing in this book). Both are *Newtonian fluids*, i.e. *viscous fluids* with bulk viscosities of $\eta = 15$ P (poise) for dOS and $\eta = 50$ P for OEP. The contact angle[2] θ that one drop of liquid polymer has on the other one is about 8°, hence very small (the liquids wet each other almost completely). The liquid polymers were studied at a concentration ratio of 0.5, and films of the mixture were put on gold-coated silicon wafers as substrates. The films had initial thicknesses between 80 and 800 nm.

Figure 1.3 shows the course of a typical experiment with this system in a schematic representation. Starting out with a film of the mixture placed on a silicon wafer (Fig. 1.3(a)), two observations are made. A thin liquid layer ruptures from a thick liquid layer on top of the wafer substrate rather than from the substrate itself: the liquid has first demixed (Fig. 1.3(c)) and only then dewets (Fig. 1.3(c)). It is found in the experiment that the film dewets from the substrate starting from the sample edge inwards and, upon its retraction, leaves droplets behind. Dewetting is brought about by the formation of holes at the film edge which is caused by fluctuations of both the liquid-liquid and liquid-gas interfaces (Fig. 1.3(c)).

Several factors need to be taken into account to explain these experiments. These are:

- The surface tensions between the liquids, the vapour and the substrate. The situation is more complex than the previous one due to the binary mixture: apart from having two components, this can also lead to concentration gradients across the sample, which in turn imply surface tension gradients with an associated mass transport via the so-called *Marangoni flow*;
- Demixing creates a *sandwich structure*: the gold-covered silicon wafer with two liquid layers on top of each other rather than one on the solid support. Thus, one has a system with three interfaces (solid-liquid, liquid-liquid, and liquid-gas) with the latter two even interacting dynamically;
- The sample boundary at which dewetting is initiated.

A full theory that can quantitatively cover all these different facts has not yet been formulated; in view of the different aspects this system contains, this is indeed a formidable—but not impossible—task.

---

[2] See Chap. 2 for a precise definition of the contact angle.

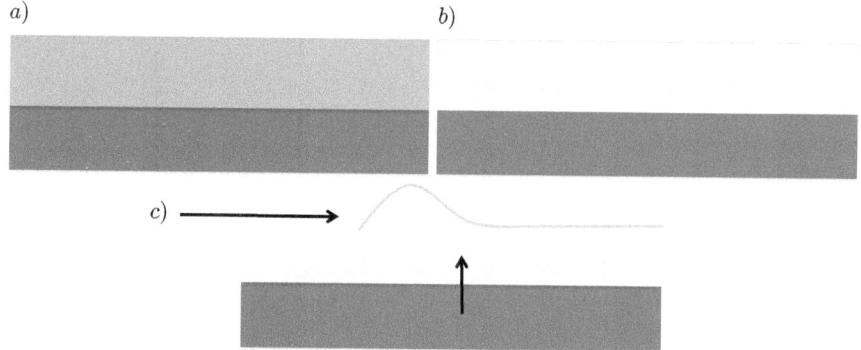

Fig. 1.3 Dewetting of a liquid film from a binary mixture in a schematic drawing: (**a**) a film of the mixture placed on top of a silicon wafer; (**b**) the film has demixed into its two components; (**c**) rupture of the thin film on top of the bottom layer: a perturbation from the film edge propagates inward, as indicated by the horizontal arrow. The vertical arrow points to a thinning zone at which droplets will start to form and disconnect from the film. Drawn after (Yerushalmi-Rozen et al. 1999 and, in particular, Kerle et al. 2002, notably Fig. 10 in that paper)

These two examples shall suffice us as an appetizer. While the first convinces us that a quantitative accord between experiment and theory is possible, the second shows us that we have to limit ourselves to well-controllable factors if we want to achieve a quantitative, mathematical description. More features render the systems more complex and will quickly require an extension of our theoretical approach.

In this book, we will explain the theoretical concepts needed to understand and quantitatively model wetting and dewetting phenomena and polymer flow. The following second chapter covers the theoretical fundamentals of the behaviour of liquids at surfaces, the physics of wetting and dewetting phenomena. Here, the basic notions of surface tensions, contact angles and dispersion forces will be introduced and explained, as well as the main theoretical tool of this part of the book: the *effective interface Hamiltonian*. Chapter 3 then confronts these concepts to experimental reality.

The subsequent chapters four and five in Part II of the book cover hydrodynamic and viscoelastic properties of thin polymeric films. For polymers, one can modify the interaction between the polymeric film and the substrate, and in this way change the friction (or 'slip') of the film on the surface. This alone is already a more complex topic, but things turn really complicated when one has to take into account as well the thermodynamic state of the film, which for a solid or near-solid film will show characteristics of a glass. Here we arrive at the limit of what is currently really understood for these systems.

All in all, the ambition of this book is to provide depth and a certain breadth in the topics covered. This means that by way of compromise, some aspects are treated in more detail while others are only sketched. To learn more about these other cases, the reader will be asked to consult additional literature that will be referred to in the conclusion chapter (Chap. 6) of this book.

Chapter 2
Statistical Mechanics of Thin Films

The second chapter is an introduction into the main physical concepts of wetting and dewetting, in particular to the corresponding *wetting* and *dewetting phase transitions*. These phase transitions occur in a broad range of physical systems: classical liquids, polymers, quantum liquids such as helium... and even superconductors, in which the 'liquid' phases are provided by the electrons, both normal and superconducting. We will encounter some of these physical examples in Chap. 3.

In this and the following chapters a number of calculational exercises, called *Tasks*, are indicated. They are meant to induce the reader to practice and deepen the understanding of the developed concepts. Occasionally, these tasks will be more than simple completions of the derivations performed in the book.

2.1 Capillarity and Surface Tension

The basic equation governing capillary phenomena at surfaces is *Young's equation* (also called the *Young-Dupré equation*) which relates the interfacial tensions between a solid s, liquid l and vapour phase v denoted by σ_{ij} with $i, j = s, l, v$ via the *contact angle* which the liquid-vapour interface makes with the solid wall. The equation reads as

$$\cos\theta = \frac{\sigma_{sv} - \sigma_{sl}}{\sigma_{lv}} = 1 + \frac{S}{\sigma_{lv}} \tag{2.1}$$

where $S \equiv \sigma_{sv} - (\sigma_{sl} + \sigma_{lv})$ is called the *spreading coefficient*.

> *Task: Show that in thermal equilibrium, $S < 0$ while $S > 0$ is possible in a metastable or unstable state. This terminology refers to the fact that if $S = 0$, a droplet has fully spread to cover the available surface: its equilibrium contact angle has gone to zero, the surface is hence wet.*

Young's equation follows from a simple argument which is illustrated in Fig. 2.1. Interfacial tensions are, by physical dimension, free energies per unit area, hence forces, and therefore Young's equation follows from a *force balance* or a *condition*

R. Blossey, *Thin Liquid Films*, Theoretical and Mathematical Physics, DOI 10.1007/978-94-007-4455-4_2, © Springer Science+Business Media Dordrecht 2012

Fig. 2.1 Mechanical
equilibrium condition leading
to Young's equation: surface
free energies σ_{ij} with
$ij = (sl, lv, sv)$ at the
three-phase contact line of a
liquid drop placed on a solid
substrate. Drop and substrate
are shown in different *gray
shades*

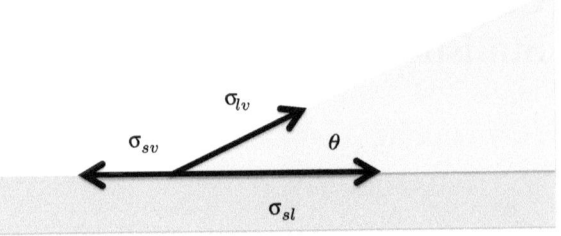

of mechanical equilibrium at the points where the three phases meet, the *three-phase contact line*. This situation applies to both a capillary filled with liquid (such as encountered in the Introduction), or to a droplet placed on a flat wall, as sketched in Fig. 2.1 (actually, only a cut through the foot of the drop is shown).

Let us consider the droplet case in some detail. We ascribe to the liquid-vapour interface with surface tension σ_{lv} at a height $h(x)$ the *interface Hamiltonian*

$$\mathcal{H}[h] = \int d^2x \left[\sigma_{lv} \left(\sqrt{1 + \left(\nabla h(x)\right)^2} - 1 \right) - (\Delta\mu)h(x) \right] \tag{2.2}$$

where the first term under the integral corresponds to the surface free energy of the liquid-vapour interface measured relative to a flat interface and written in a *Monge representation* which assumes that there is no overhang in the interface, see Fig. 2.2. The second term describes the chemical potential difference between the two fluid phases (vapour/liquid). The parameter $\Delta\mu$ can also be read as a Lagrange multiplier associated with a fixed droplet volume, $\Omega \equiv \int d^2x h(x)$.

We further assume that the lateral dimension of the drop will be large as compared to its height so that the gradients of h will be small; we can then linearize the surface term according to

$$\sqrt{1 + \left(\nabla h(x)\right)^2} \approx 1 + (1/2)\left(\nabla h(x)\right)^2. \tag{2.3}$$

Finally, we assume a cylindrical symmetry of the interface profile, $h \equiv h(r)$, with r as the radial coordinate. We can now calculate the first variation of $\mathcal{H}$ and find the ODE

$$\sigma_{lv}\left(h''(r) + \frac{1}{r}h'(r) \right) - \Delta\mu = 0, \tag{2.4}$$

where the prime stands for the differentiation with respect to r. This equation is the (linearized) *Young-Laplace equation*.

In order to obtain a droplet profile from Eq. (2.4) we impose the two boundary conditions

$$h'(0) = 0, \qquad h(R) = 0. \tag{2.5}$$

The solution one obtains for the droplet profile is given by the parabola

$$h(r) = H\left(1 - (r/R)^2\right) \tag{2.6}$$

where one has the two parameters, the droplet height H and its radius R, given by

$$H = 2|S|(\Delta\mu)^{-1}, \qquad R = 2\sqrt{2\sigma_{lv}|S|}(\Delta\mu)^{-1}. \tag{2.7}$$

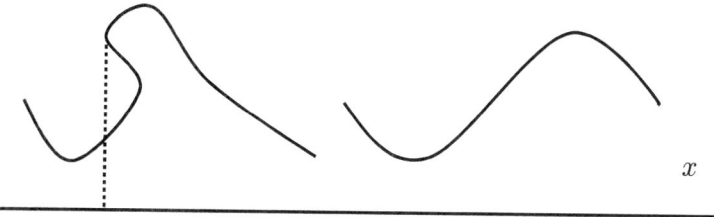

Fig. 2.2 Sketch of two interface configurations, one with and one without an overhang. The overhanging configuration does not have a unique height function $h(x)$, as indicated by the dashed vertical line

Task: Do this calculation explicitly.

In our 'parabolic' approximation—having simplified the interface Hamiltonian by just keeping the harmonic term—the spreading coefficient S is given by

$$|S| = \frac{\sigma_{lv}\theta^2}{2} \tag{2.8}$$

which follows from an expansion of the cosine in Young's equation. The ratio of height and width of the droplet is obtained from Eqs. (2.7)

$$\frac{H}{R} = \sqrt{\frac{|S|}{2\sigma_{lv}}} \tag{2.9}$$

and hence is a linear function of contact angle. One can also easily calculate the surface free energy of the sessile droplet which simply is

$$\mathcal{E} \equiv \mathcal{H}[h(r)] = \pi |S| R^2. \tag{2.10}$$

These results apply to droplets of sizes for which gravitational forces do not play a role. We have already seen in the Introduction that the relevant length here is the *capillary length*

$$\ell_c \equiv \sqrt{\frac{\sigma_{lv}}{\varrho g}}, \tag{2.11}$$

such that in our case we must have $H < \ell_c$ for the theory to be applicable (in addition to the condition on the contact angle, or on the gradients of h).

Task: Calculate the variation of the interface Hamiltonian without the linearization for small gradients, i.e. by using the full Hamiltonian, Eq. (2.2). Determine the full droplet profile and its surface free energy.

Can we validate this result experimentally to convince ourselves that it is a valid description of the droplet properties? The challenge is, of course, to look at small droplets with dimensions well below the capillary length such that gravity plays no role.

In order to achieve this, polymer droplets of PS (polystyrene) have been studied with *Atomic Force Microscopy*, AFM (Seemann et al. 2001a). Contact angles θ

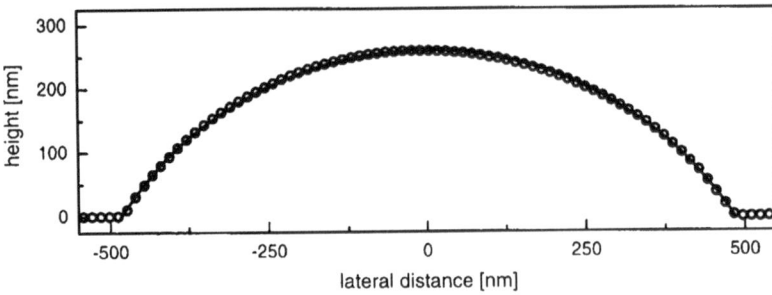

Fig. 2.3 An AFM scan of a sessile droplet profile and its fit to a parabola. Reprinted with permission from Seemann et al. (2001a). Copyright by IOP

were determined for two different substrates: SiO-wafers and OTS-wafers. The experimental methods to produce such droplets (and, particularly, thin polymer films) are described in Appendix A. Two methods were used for analysis: (i) the determination of the slope at the three-phase contact line and (ii) a fit of a spherical cap to the data. The latter result is shown in Fig. 2.3.

Figure 2.4 shows the ratio of the scaling parameter H/R as a function of contact angle. Two clusters of data are shown, characterizing the two substrates. Droplets on SiO-wafers have a contact angle of 6.9(5)° with a height of about 20–40 nm. Droplets on OTS-wafers have a larger contact angle of 58(1)° and a central height of about 200–550 nm. The solid line in Fig. 2.4 corresponds to the parabolic droplet, the dashed line to a spherical droplet. These results convincingly show that the model description covers the shape of 'small' droplets very well, and that for larger contact angles indeed the full spherical shape needs to be taken into account. This reasoning applies to the 'macroscopic' shape of the drop.

Things are different, however, when one looks into the details of the three-phase contact line. Indeed one might expect that additional forces, not covered by our interface Hamiltonian, may lead to modifications of the simple picture. The contact-line is placed in a region where more microscopic details of the interactions may matter, and the question therefore is how to capture these effects, beyond Young's equation.

Figure 2.5 plots the cosine of the contact angle against the curvature of the interface. Underlying this plot is the assumption that the contact angle obeys a *modified Young equation* of the form

$$\cos\theta_r = \cos\theta - \frac{\tau\kappa}{\sigma_{lv}} \tag{2.12}$$

where κ is the *interfacial curvature*,[1] which in this case is simply $1/R$, and τ a 'line tension' associated with the droplet contact line. This equation covers the deviation of the interfacial profile next to the wall, due to forces not included into the Hamiltonian of a capillary interface.

---

[1] See Chap. 4 for an explicit mathematical expression for the interfacial curvature.

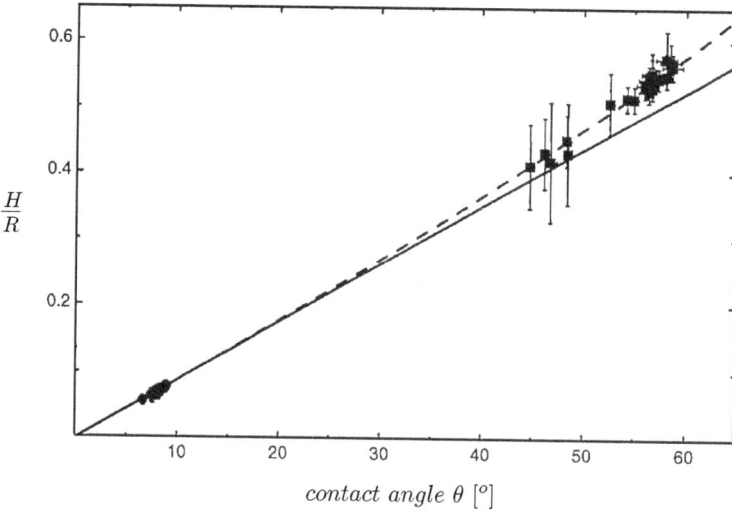

Fig. 2.4 The scaling of the droplet shape: H/R as a function of contact angle. Reprinted with permission from Seemann et al. (2001a). Copyright by IOP

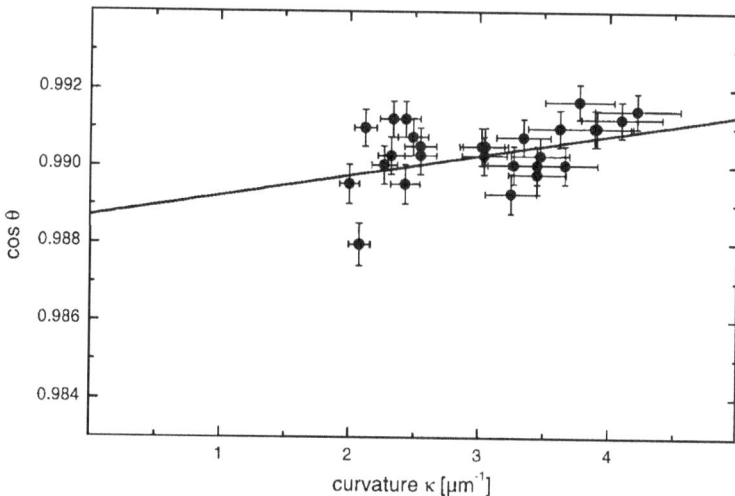

Fig. 2.5 The line tension of a polystyrene droplet on an SiO-wafer, assuming the modified Young equation (*straight line*). Reprinted with permission from Seemann et al. (2001a). Copyright by IOP

From Fig. 2.5 we can infer a value of this 'line tension' of $\tau = -10^{-11}$ J/m. Since a typical value of the surface tension is on the order of $\sigma_{lv} = 10^{-2}$ J/m^2, the ratio $|\tau/\sigma_{lv}|$ gives a characteristic length scale on which the effect of the line tension matters, which is 1 nm. This result clearly shows the challenge inherent to such measurements.

Fig. 2.6 SFM topography image of a water droplet on a silicon surface with a striped pattern with different wettability properties. The middle region is hydrophobic, causing the contact line to bend inwards. Reprinted with permission from Pompe and Herminghaus (2000). Copyright by the American Physical Society

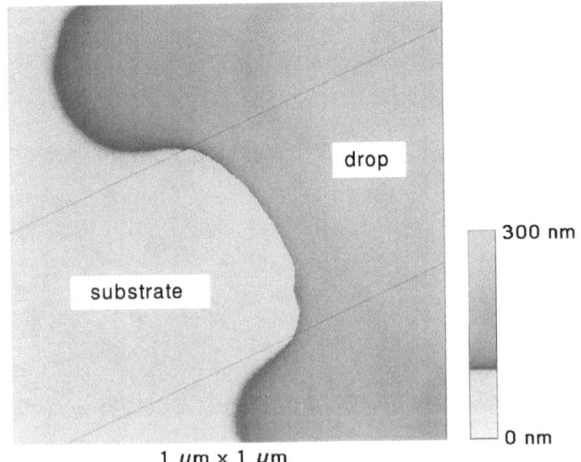

Fig. 2.7 The vapour-liquid interface profile, here called $\ell(x)$, measured by SFM for three different systems. Reprinted with permission from Pompe and Herminghaus (2000). Copyright by the American Physical Society

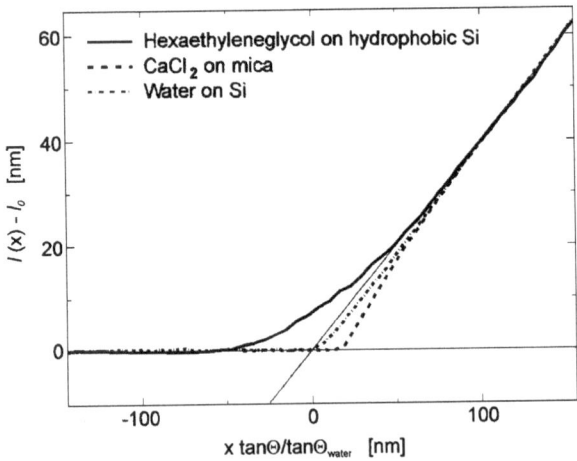

More details of the profile of the contact line were described by Hermingaus et al. (1999) and Pompe and Herminghaus (2000) who created liquid surface topographies based on artificially striped wettability patterns which force the contact-line of the drop to bend strongly, see Fig. 2.6. Surface force measurements not only allow an analogous determination of τ as before with values via the modified Young equation, and yielding line tension values of similar magnitude, but they can also resolve the details of the vapour-liquid interface. Figure 2.7 shows these profiles for three different surfaces. All data show deviations from the straight ('*dividing*') line which corresponds to the droplet shape obtained from the Young-Laplace equation. The observation therefore quantifies the surface force contributions not taken into account by the surface tension, and it is precisely these forces which control the behaviour of the line tension, and, as we will see, also of thin films.

In order to achieve an understanding of the line tension in a proper theoretical framework—e.g., of its sign—we must first look into more detail at the origin of the microscopic forces acting at interfaces. We will see that we have and how we have to correct the *interface Hamiltonian* by additional surface forces. Although we have understood from the previous discussion that there is a curvature-related correction to Young's equation, and that the profile of the droplet foot deviates from a simple straight line, we do not yet have a clear understanding what the 'line tension' really is, and, actually, whether both measurements have access to the same quantity. Measurements alone do not suffice; conceptual, hence theoretical understanding is needed here.

2.2 Forces Acting at Interfaces

The *surface tensions* (or *surface forces*) take into account the macroscopic effects of intermolecular interactions which can well be assumed as being carried by short-range, structural interactions. However, on the molecular level interactions are not only structural, but also determined by charges, hence of an electrostatic nature. The fundamental theory of *dispersion forces* goes back to Dzyaloshinskii, Lifshitz and Pitaevski (short: DLP) (Dzyaloshinskii et al. 1961). The quantity of interest describing the attraction or repulsion of two interfaces as sketched in Fig. 2.8 is given by the thermodynamic expression of the *disjoining pressure*

$$\Pi_{vdW}(h) \equiv -\frac{1}{A}\left(\frac{\partial G_{film}}{\partial h}\right) = -\frac{1}{A}\left(\frac{\partial V}{\partial h}\right), \qquad (2.13)$$

where A is surface area, and $V(h)$ is the Gibbs free energy of the two-surface system which is also called the *effective interface potential*. $V(h)$ is understood as the excess surface free energy per unit area it takes to bring two separated interfaces (hence located 'at infinity') together to a finite distance h.

DLP showed how Π_{vdW} can be determined based on the theory of *finite-temperature quantum electrodynamics*. They derived the fundamental expression for the *disjoining pressure* $\Pi_{vdW}(h)$ for a planar, film-type geometry shown in Fig. 2.8.

Fig. 2.8 Geometry for the calculation of the long-range force between dielectric media across a planar gap

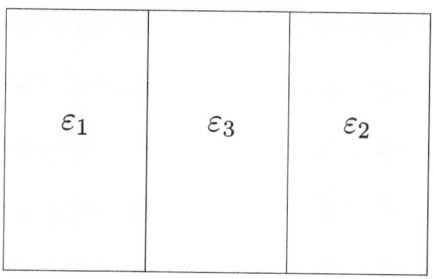

The three media (solid, liquid film, vapour) are characterized by their frequency-dependent dielectric functions ε_i ($i = 1, 2, 3$). The final result writes as[2]

$$\Pi_{vdW}(h) = -\frac{k_B T}{\pi c^3} \sum_{n=0}^{\infty} \varepsilon_3^{3/2} \zeta_n^2 \int_1^{\infty} dp\, p^2 [I_1(h, p\zeta_n) + I_2(h, p\zeta_n)] \qquad (2.14)$$

where $\zeta_n = 2\pi n k_B T / \hbar$ and the complex dispersion functions $\varepsilon_j(i\zeta_n)$ are introduced which are related to the frequency-dependent dielectric functions via *Kramers-Kronig relations*. In this expression, one has

$$I_1(h, p\zeta_n) = \left(\Delta_1(p)\Delta_2(p) \exp\left(\frac{2p\zeta_n}{c} h\sqrt{\varepsilon_3} - 1 \right) \right)^{-1} \qquad (2.15)$$

and

$$I_2(h, p\zeta_n) = \left(\Delta_{13}(p)\Delta_{23}(p) \exp\left(\frac{2p\zeta_n}{c} h\sqrt{\varepsilon_3} - 1 \right) \right)^{-1} \qquad (2.16)$$

with

$$\Delta_j(p) = \frac{s_j - p}{s_j + p} \qquad (2.17)$$

$$\Delta_{jk}(p) = \frac{s_j \varepsilon_k - p\varepsilon_j}{s_i \varepsilon_k + p\varepsilon_j} \qquad (2.18)$$

with

$$s_j = \left(\varepsilon_j / \varepsilon_3 - 1 + p^2 \right)^{1/2}. \qquad (2.19)$$

The DLP-formula allows to distinguish two contributions:

$$\Pi_{vdW}(h) = \Pi_{n=0}(h) + \Pi_{n>0}(h), \qquad (2.20)$$

i.e., the zero-frequency and higher-order contributions. To make the first term explicit, one can first replace the integration over the variable p by an integration over

$$x = \frac{2p\zeta_n h \varepsilon_3^{1/2}}{c} \qquad (2.21)$$

to obtain

$$\Pi_{vdW}(h) = -\frac{k_B T}{8\pi h^3} \sum_{n=0}^{\infty} \int_{x(\zeta_n)}^{\infty} dx\, x^2 \left[\left(\frac{1}{\Delta_1 \Delta_2} e^x - 1 \right)^{-1} + \left(\frac{1}{\Delta_{13} \Delta_{23}} e^x - 1 \right)^{-1} \right] \qquad (2.22)$$

with the lower limit of integration being given by

$$x(\zeta_n) = 2\zeta_n h \varepsilon_3^{1/2} / c \qquad (2.23)$$

---

[2] With respect to the usual notational difficulty, note that film height h should not be confused with Planck's (reduced) constant $\hbar$.

and the functions Δ_j and Δ_{jk} rewrite as

$$\Delta_j = \frac{(4\zeta_n^2 h^2 (\varepsilon_j - \varepsilon_3) + (cx)^2)^{1/2} - cx}{(4\zeta_n^2 h^2 (\varepsilon_j - \varepsilon_3) + (cx)^2)^{1/2} + cx} \tag{2.24}$$

and

$$\Delta_{jk} = \frac{(4\zeta_n^2 h^2 (\varepsilon_j - \varepsilon_3) + (cx)^2)^{1/2} - cx\varepsilon_j/\varepsilon_k}{(4\zeta_n^2 h^2 (\varepsilon_j - \varepsilon_3) + (cx)^2)^{1/2} + cx\varepsilon_j/\varepsilon_k}. \tag{2.25}$$

For the first term of $\Pi_{vdW}(h)$, one has $\Delta_1 = \Delta_2 = 0$ and the term simplifies to

$$\Pi_{n=0}(h) = -\frac{k_B T}{16\pi h^3} \int_0^\infty dx\, x^2 \left(\frac{(\varepsilon_3(0) + \varepsilon_1(0))(\varepsilon_3(0) + \varepsilon_2(0))}{(\varepsilon_3(0) - \varepsilon_1(0))(\varepsilon_3(0) - \varepsilon_2(0))} e^x - 1 \right)^{-1} \tag{2.26}$$

where $\varepsilon_j(0)$ are the *static dielectric constants* of the three media $j = 1, 2, 3$. This can be finally approximated by neglecting the (-1) term in the bracketed expression since the integral is dominated by the first term. After completing the integral one concludes with

$$\Pi_{n=0}(h) = -\frac{k_B T}{8\pi h^3} \int_0^\infty dx\, x^2 \frac{(\varepsilon_3(0) + \varepsilon_1(0))(\varepsilon_3(0) + \varepsilon_2(0))}{(\varepsilon_3(0) - \varepsilon_1(0))(\varepsilon_3(0) - \varepsilon_2(0))}. \tag{2.27}$$

This first, static term of the disjoining pressure represents the contribution of the permanent dipolar orientations of the materials, usually named the *Keesom* and *Debye* *contributions*.

The second, dispersive term can be rewritten by noting that because of the exponential, only contributions for which $\zeta_n \gg c/a$ contribute to the final result. We therefore replace the sum over frequencies by an integral over the frequency range $d\zeta = (2\pi k_B T/\hbar)dn$ which works for $hk_B T/c\hbar \ll 1$. Around room temperature this restriction means that film thicknesses should be smaller than 10^4 Å, which is generally the case. One thus finds

$$\Pi_{vdW}(h) = -\frac{k_B T}{16\pi h^3} \int_{\zeta_1}^\infty d\zeta \int_{x(\zeta_n)}^\infty dx\, x^2 \left[\left(\frac{1}{\Delta_1 \Delta_2} e^x - 1 \right)^{-1} \right.$$
$$\left. + \left(\frac{1}{\Delta_{13}\Delta_{23}} e^x - 1 \right)^{-1} \right]. \tag{2.28}$$

To work further with this result one can distinguish again two thickness regimes. They derive from electrostatic retardation effects. Comparing wavelengths to film thicknesses, one finds that for lengths less than 500 Å, retardation effects can be neglected. Going through the dominant contributions in the integrand fir these length scales, one finds $\Delta_1 \approx \Delta_2 \approx 0$ and obtains for the $(n > 0)$-part of the dispersion forces

$$\Pi_{n>0}(h) = -\frac{k_B T}{16\pi h^3} \int_{\zeta_1}^\infty d\zeta \int_0^\infty dx\, x^2 \left(\frac{(\varepsilon_3 + \varepsilon_1)(\varepsilon_3 + \varepsilon_2)}{(\varepsilon_3 - \varepsilon_1)(\varepsilon_3 - \varepsilon_2)} e^x - 1 \right)^{-1} \tag{2.29}$$

where $\varepsilon_j = \varepsilon_j(i\zeta)$ for $j = 1, 2, 3$. Similarly as before we can further reduce the integral and perform the integration over x to end up with

$$\Pi_{n>0}(h) = -\frac{k_B T}{8\pi h^3} \int_{\zeta_1}^\infty d\zeta \frac{(\varepsilon_3(i\zeta) + \varepsilon_1(i\zeta))(\varepsilon_3(i\zeta) + \varepsilon_2(i\zeta))}{(\varepsilon_3(i\zeta) - \varepsilon_1(i\zeta))(\varepsilon_3(i\zeta) - \varepsilon_2(i\zeta))} \tag{2.30}$$

For the case of large film thicknesses one introduces a new variable $y = 2pl\zeta/c$ to obtain the expression

$$
\Pi_{n>0}(h) = -\frac{\hbar}{32\pi^2 h^4} \int_0^\infty dy \int_1^\infty dp \frac{y^2}{p^2} \varepsilon_3^{3/2}
$$
$$
\times \left[\left(\frac{1}{\Delta_1(p)\Delta_2(p)} e^{x\sqrt{\varepsilon_3}} - 1 \right)^{-1} \right.
$$
$$
\left. + \left(\frac{1}{\Delta_{13}(p)\Delta_{23}(p)} e^{x\sqrt{\varepsilon_3}} - 1 \right)^{-1} \right]
\tag{2.31}
$$

This expression shows that the retarded contribution falls off as h^{-4}.

From this excursion we now finally retain that the dispersion forces can be described by a disjoining pressure

$$
\Pi_{vdW}(h) = -\frac{A}{6\pi h^3}
\tag{2.32}
$$

with the *Hamaker constant*

$$
A \equiv A_{n=0} + A_{n>0}
\tag{2.33}
$$

or

$$
A \approx \frac{3}{4}kT \left(\frac{\varepsilon_1 - \varepsilon_3}{\varepsilon_1 + \varepsilon_3} \right) \left(\frac{\varepsilon_2 - \varepsilon_3}{\varepsilon_2 + \varepsilon_3} \right)
$$
$$
+ \frac{3\hbar}{4\pi} \int_{\zeta_1}^\infty dv \left(\frac{\varepsilon_1(i\zeta) - \varepsilon_3(i\zeta)}{\varepsilon_1(i\zeta) + \varepsilon_3(i\zeta)} \right) \left(\frac{\varepsilon_2(i\zeta) - \varepsilon_3(i\zeta)}{\varepsilon_2(i\zeta) + \varepsilon_3(i\zeta)} \right)
\tag{2.34}
$$

Since the frequency ζ_1 is usually large against contributions from molecular rotations, the dispersion term is mostly determined by electronic contributions. With a simple model for the electronic dielectric behaviour

$$
\varepsilon(i\zeta) = 1 + \frac{n^2 - 1}{1 + \zeta^2/\zeta_e^2}
\tag{2.35}
$$

for each medium, where n is the index of refraction, the integration over the frequencies can be performed and one obtains a simplified formula containing only macroscopic quantities. It reads as

$$
A = A_{n=0} + A_{n>0} \approx \frac{3}{4}kT \left(\frac{\varepsilon_1 - \varepsilon_3}{\varepsilon_1 + \varepsilon_3} \right) \left(\frac{\varepsilon_2 - \varepsilon_3}{\varepsilon_2 + \varepsilon_3} \right) + \frac{3\hbar\zeta_e}{8\sqrt{2}} F(n_1, n_2, n_3)
\tag{2.36}
$$

where

$$
F(n_1, n_2, n_3) \equiv \frac{(n_1^2 - n_3^2)(n_2^2 - n_3^2)}{(n_1^2 + n_3^2)^{1/2}(n_2^2 + n_3^2)^{1/2}((n_1^2 + n_3^2)^{1/2} + (n_2^2 + n_3^2)^{1/2})}.
\tag{2.37}
$$

This formula shows that it is possible to tune the surface interactions by choosing surfaces with adequate dielectric behaviour and refractive indices such that the interaction terms, as a function of temperature, can show interesting behaviour: in particular a change of sign, which then signals a change from favorable to unfavorable interactions, hallmark of a transition between two different states, favoring either wetting or dewetting.

A classification of the different types of forces that are ultimately of electrostatic origin, and their dependence on geometry, is given in the book by Israelachvili (1992).

2.3 Wetting and Dewetting: The Wetting Phase Diagram

The insights gained from the previous section now allow us to formulate the basic concepts of wetting beyond Young's equation, and in particular to address the notion of a *wetting (or dewetting) phase transition*. The basic mathematical object in this context is the *effective interface potential* $V(h)$. The notion *effective* in this context refers to the fact that this potential is not a microscopic potential, but that its parameters depend on temperature—e.g., via their indices of refraction, as shown in the previous section.

For large separations, $h \to \infty$, $V(h) \to 0$, and in what follows we assume for the asymptotics for large separations a van der Waals interaction of the form

$$V(h) \sim \frac{A}{h^{m-1}}, \quad h \gg 0 \tag{2.38}$$

where we have collected the numerical factors into a redefined *Hamaker constant* A which we will use from now on. We call the exponent m in order to cover more general cases than the *non-retarded van der Waals forces*, $m = 3$, or the *retarded* van der Waals forces with $m = 4$; also other values $m > 1$ are possible. Even the (formal) case of $m \to \infty$ is possible which we identify with exponentially decaying forces

$$V(h) \sim \exp(-h/\xi) \tag{2.39}$$

with a decay length ξ. These forces correspond to short-ranged interactions between interfaces. The wetting phase transition was originally described in terms of the Ginzburg-Landau theory of phase transitions (Cahn 1977; Nakanishi and Fischer 1982) to which such forces apply.

Figure 2.9 shows three possible shapes the interface potential $V(h)$ can display in the opposite limit $h \to 0$, i.e. close to the wall at which structural effects intervene (steric repulsion). These cases are

- a *stable thin film*: the force needed to push the two interfaces together continues to increase upon approach of the interfaces;
- an *unstable film*: the interfaces feel a strong attraction until a minimum is reached below which the force increases again. Note that, in this case, the Hamaker constant $A < 0$;
- a *metastable film*: upon the approach of the two interfaces, the force first increases and then decreases before reaching a minimum. This is the typical behaviour

Fig. 2.9 Schematic drawing of the effective interface potential $V(h)$. Three cases are depicted: *short dashes*: stable thin film; a thick film on this substrate would be unstable; *long dashes*: stable thick film; *solid line*: barrier separating a thin and a thick film state, with the relative stability being determined by the height difference between the two minima. In the case shown, the thin film is stable since $S = V(h_{min}) < 0$

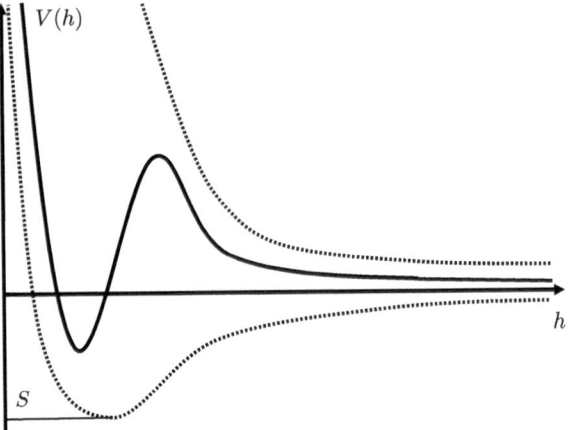

expected when a free energy barrier needs to be surmounted. The relative position of the two minima in $V(h)$ decides which is the metastable and which the stable state; in the potential shown, a microscopically thin film is stable on the surface, or an infinitely thick film.

The effective interface potential shown concerns the situation at chemical equilibrium, for which the chemical potential between the fluid phases is zero: $\Delta\mu = 0$. This, of course, need not be the case and we can therefore consider the more general full effective potential

$$\Phi(h) \equiv V(h) - (\Delta\mu)h. \qquad (2.40)$$

$V(h)$ and $\Phi(h)$ are compared for the case of a metastable state, again for the case for a metastable thick film in Fig. 2.10. The left figure shows a situation at coexistence with a stable thin film.

Figure 2.10 (right) displays the effective interface potential $\Phi(h)$ in a situation in which a thick film is metastable on the wall. In the full interface potential $\Phi(h)$ the minimum of $V(h)$ at $h = \infty$ is, for $\Delta\mu < 0$, shifted to a finite value h_1. Along a prewetting line the two minima of $\Phi(h)$ have equal height, and they coincide at the prewetting critical point. The prewetting line is therefore the coexistence line of prewet and nonwet states.

For the case of an interface potential with a barrier, for which either the microscopically thin or the thick film state can be the metastable state, the corresponding thermodynamic states can be characterized by the generic wetting phase diagram shown in Fig. 2.11. This diagram fundamentally underlies all cases treated in this book; we will see in Chap. 3 different realizations in different physical systems. We denote the prewetting line in this text interchangeably as $T_p(\Delta\mu)$ or, inversely, as $\Delta\mu_p(T)$, or $\Delta\mu_p(S)$. The spreading coefficient can be expressed in terms of temperature distance from the wetting transition W.

Figure 2.11 shows the phase diagram for a system which can have *first-order wetting transition* in terms of temperature T and chemical potential μ. Above the

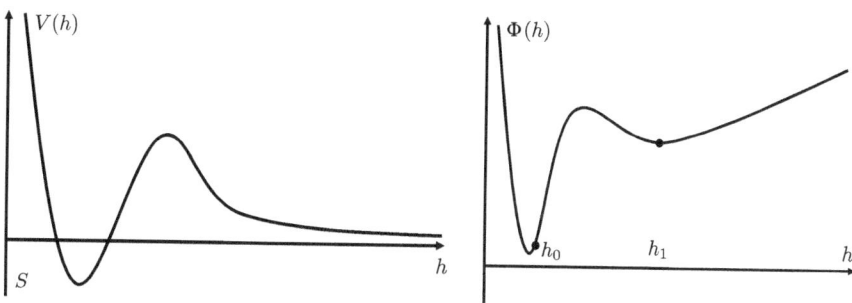

Fig. 2.10 *Left*: The effective potential $V(h)$, where $S \equiv V(h_{min})$ with the absolute minimum of V located at h_{min}; *right*: the full potential $\Phi(h) \equiv V(h) - (\Delta\mu)h$. Here, h_1 is the equilibrium thickness of the undercooled layer, and $h_1 - h_0$ is the depth of the critical hole

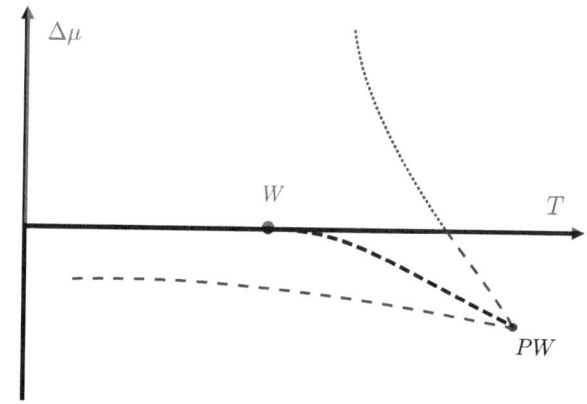

Fig. 2.11 Generic wetting-dewetting phase diagram in the temperature-chemical potential plane. Indicated are the upper and lower surface spinodals (*thin dashed* and *dotted*), the first-order wetting transition point (W), the prewetting line (*thick dashed*) and the prewetting critical point (PW). The region above the coexistence curve is the bulk liquid side of the phase diagram (shown *shaded in gray*). The surface spinodal on the liquid side is indicated as a *dotted line* since the bulk liquid phase is stable on this side of the phase diagram

coexistence value $\mu = \mu_c$ of the two-fluid bulk system the thickness of a wetting layer is infinite whereas for $\Delta\mu \equiv \mu - \mu_c < 0$ it is finite. In the limit $\Delta\mu \to 0$ from below the layer thickness diverges continuously above the wetting temperature T_w at W, but it has an infinite jump across the partial-wetting line $T < T_w$, $\Delta\mu = 0$, reflecting the instability of the bulk phase. A finite jump from a thin to a thick layer occurs when the prewetting line $T_p(\Delta\mu)$ is crossed from the region $T < T_p(\Delta\mu)$ to $T > T_p(\Delta\mu)$. This jump runs to infinity when T_w is approached along the prewetting line, and it disappears at the prewetting critical point PW.

By contrast, for a system without a barrier, there is always only one minimum at $\Delta\mu = 0$ which is either at a small, finite value of h, or at infinity. The phase transition between these two states occurs at a value $T = T_c$, $\Delta\mu = 0$ and is called a *critical wetting transition.* It corresponds to the limit in which the prewetting critical point *PW* of Fig. 2.11 coincides with the wetting transition *W* at $T = T_w$. We will only briefly touch on this case in Chap. 3, but otherwise stay entirely away from the critical wetting case (which has a rich history in its treatment of fluctuation effects).

2.4 The Hamiltonian and the Line Tension

2.4.1 The Effective Interface Hamiltonian

We now have the two essential ingredients for a full mathematical interface model: the surface free energy from capillary theory, and the dispersion forces for 'microscopic'—thin—films. We collect them together into the *effective interface Hamiltonian*

$$\mathcal{H}[h] = \int d^{d-1}x \left[\frac{\sigma_{lv}}{2}(\sqrt{g} - 1) + V(h) - (\Delta\mu)h \right] \qquad (2.41)$$

where $\sqrt{g} = (1 + (\nabla h)^2)^{1/2}$. We will mostly be interested in inhomogeneous states of the system (droplets and holes) and therefore have to study the variational problem

$$\frac{\delta \mathcal{H}[h]}{\delta h(x)} = 0 \qquad (2.42)$$

which leads to a *nonlinear elliptic differential equation*

$$-\sigma_{lv}\nabla\left(\frac{\nabla h(x)}{\sqrt{g}}\right) + V'(h(x)) - \Delta\mu = 0. \qquad (2.43)$$

What remains to be specified is the symmetry of the sought solution, as well as the proper boundary conditions.

We begin this study rightaway, for the case of the line tension. Our first task is to derive the *modified Young's equation*, Eq. (2.12). We follow a discussion presented in Dobbs (1999a).

2.4.2 Line Tension I: The Modified Young Equation

We have seen before that the shape of the drop in the vicinity of the substrate deviates from the *dividing line* which is the macroscopic interpolation of the droplet profile down to the surface. Repeating the equation we have

$$\cos\theta_r = \cos\theta - \frac{\tau\kappa}{\sigma_{lv}} \qquad (2.44)$$

where θ is Young's contact angle, determined by the surface tensions. We thus see that the real (or better *apparent contact angle*) θ_r is determined by a superposition

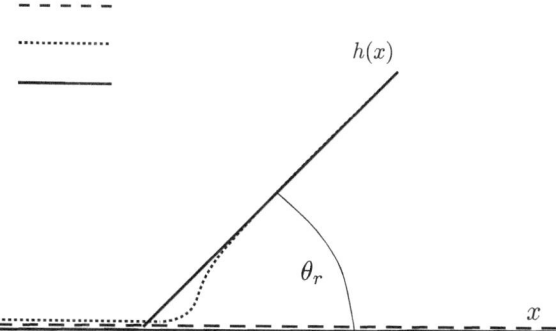

Fig. 2.12 The foot of the droplet. *Drawn line*: macroscopic approximation; *dotted line*: proper profile for an interface potential corresponding to a first-order wetting transition which asymptotically approaches a film of thickness f; *dashed line*: the thin film of thickness h_0, corresponding to the thin-film minimum of the interface potential $V(h)$. Adapted from Dobbs (1999a)

of an expression involving Young's contact angle θ and a correction due to the line tension. It is thus tempting to identify this correction term as being generated by the interface potential, and we want to see exactly how that works, based on Eq. (2.41).

Assuming cylindrical symmetry of the drop, the profile has to obey the equation (2.43) in the form

$$\sigma_{lv}\left(\frac{dh}{dr} + \frac{1}{r}\right)\left(\frac{1}{\sqrt{g}}\frac{dh}{dr}\right) = V'\big(h(x)\big) - \Delta\mu. \tag{2.45}$$

In what follows, we normalize $V_\sigma \equiv V/\sigma_{lv}$, $(\Delta\mu)_\sigma \equiv \Delta\mu/\sigma_{lv}$.

The boundary conditions to Eq. (2.45) are obviously at the center of the drop

$$h(0) = h, \qquad h'(0) = 0 \tag{2.46}$$

while at infinity we must have a merging of the profile into a thin film of thickness $f \approx h_0$.[3] We suppose conditions of partial wetting ($T < T_w$, $\Delta\mu \approx 0$) with an interface potential exactly as in Fig. 2.10. For large values $V(h)$ decays to zero. For convenience, however, we change our normalization and put the minimum of $V(h)$ at the thin film to zero: $V(h_0) = 0$; the potential then decays for large h to the value of the spreading coefficient, $-S$, noting that at partial wetting, $S < 0$. For the limit at large values of r one then has $f - h_0 = \mathcal{O}((\Delta\mu)_\sigma)$.

We have already seen that the macroscopic drop profile is determined by the chemical potential term; the Young equation part of Eq. (2.12) is thus assured. The apparent contact angle θ_r is determined from the intersection of the dividing line from the droplet profile with the substrate surface which acts as a second dividing line, see Fig. 2.12.

---

[3] Note that for one-dimensional profiles, $h_0 = h_{min}$. This is not true for $d > 1$ due to the appearance of a 'friction'-term in the ODE governing the interface profile: see the discussion in Sect. 2.5.

This identification allows to define a radius R and the contact angle θ_r which are related to the chemical potential via

$$\Delta\mu R = 2\sigma_{lv}\sin\theta_r \tag{2.47}$$

and for the height of the drop one has

$$\Delta\mu(h - f) = 2\sigma_{lv}(1 - \cos\theta_r). \tag{2.48}$$

These relations generalize the previous relations (2.4) to the present case, including the effective interface potential.

We now expand the droplet shape in powers of $1/R$ (or, more properly, ξ/R where ξ measures the radial extent on which the interface potential contributes, see Fig. 2.12). Only in the vicinity of the droplet edge the derivative of the interface potential dominates over the chemical potential difference in Eq. (2.45). To zeroth order, we therefore drop both $(\Delta\mu)_\sigma$ and the term $\sim 1/r$. The remaining equation is then identical to that of a straight contact line with a first integral (Dobbs and Indekeu 1993) given by

$$\frac{1}{\sqrt{g}}\frac{dh}{dr} \approx -\left(2V_\sigma - V_\sigma^2\right)^{1/2}. \tag{2.49}$$

Task: Verify Eq. (2.49).

Likewise, one can integrate the full equation by using the relation

$$d/dr = (dh/dr)(d/dh). \tag{2.50}$$

With the boundary conditions at small and large r one finds the equation

$$V_\sigma(h) - V_\sigma(f) = \frac{(\Delta\mu)_\sigma}{2}(h - f) + \int_f^h d\tilde{h}\left(\frac{1}{r}\frac{h}{\sqrt{g}} + \frac{(\Delta\mu)_\sigma}{2}\right). \tag{2.51}$$

For large r the solution is given by the spherical drop and the integrand then tends to zero. Obviously, as stated before, the only region of interest is around $f \approx h_0$ for which we can approximate the first term under the integral by the one-dimensional solution with $r \approx R$. Since $(f - h_0) = \mathcal{O}(1/R)$ and $V(f) = \mathcal{O}(1/R^2)$ one has, to first order:

$$V_\sigma(h) = \frac{(\Delta\mu)_\sigma}{2}(h - f) - \frac{1}{R}\int_{h_0}^h d\tilde{h}\left[\left(2V_\sigma - V_\sigma^2\right)^{1/2} - \frac{(\Delta\mu)_\sigma}{2}R\right]. \tag{2.52}$$

If the intermolecular forces decay sufficiently rapidly, the interface potential can be approximated by its asymptotic behaviour for large r, $V(h) \approx S > 0$, and the integral is convergent when the upper limit is taken to infinity. Then one obtains the modified Young equation: by using Eqs. (2.47), (2.48) and the definition of the spreading coefficient, we have

$$S = \sigma_{lv}(1 - \cos\theta_r) - \frac{\sigma_{lv}}{R}\int_{h_0}^\infty d\tilde{h}\left[\left(2V_\sigma - V_\sigma^2\right)^{1/2} - \sin\theta_r\right] \tag{2.53}$$

Fig. 2.13 The *full line* and the *symbols* show the variation of $\cos\theta_r$ with droplet radius R, numerically computed for a typical first-order wetting interface potential. The *dashed line* is the modified Young's equation with τ given by Eq. (2.55). The *insert* shows the next-to-leading order correction. Reprinted with permission from Dobbs (1999a). Copyright by World Scientific

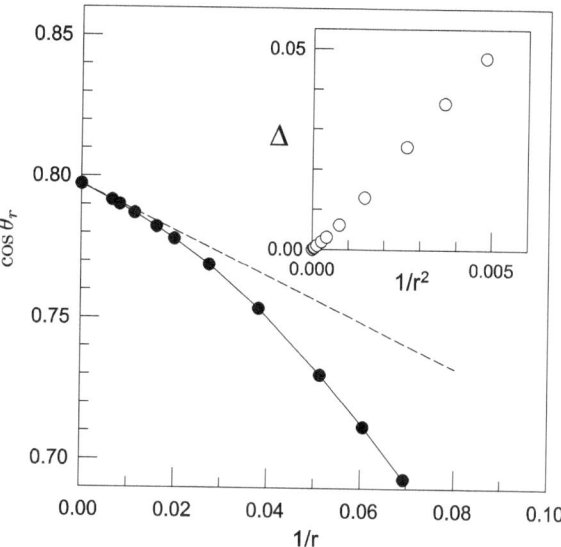

To lowest order, we replace the $\sin\theta_r$-term by $\sin\theta$ and then obtain the *modified Young equation*

$$\cos\theta_r = \cos\theta - \frac{\tau}{\sigma_{lv}R} \tag{2.54}$$

with

$$\tau = \sqrt{2}\sigma_{lv}\int_{h_0}^{\infty} d\tilde{h}\left[\left(V_\sigma - \frac{V_\sigma^2}{2}\right)^{1/2} - \left((-S/\sigma_{lv}) - \frac{(-S/\sigma_{lv})^2}{2}\right)^{1/2}\right], \tag{2.55}$$

where we now use the non-renormalized expression for V. In the case of the squared-gradient approximation, this equation simplifies to

$$\tau = \sqrt{2}\sigma_{lv}\int_{h_0}^{\infty} d\tilde{h}\left[V_\sigma^{1/2} - (-S/\sigma_{lv})^{1/2}\right]. \tag{2.56}$$

In Fig. 2.13, the variation of $\cos\theta_r$ with radius R is shown from a numerical calculation making use of an exemplary interface potential corresponding to a first-order wetting case

$$V_\sigma(h) = Ae^{-(h-1)} + Be^{-2(h-1)} + Ce^{-3(h-1)} - S/\sigma_{lv} \tag{2.57}$$

with $A = 3.3$, $B = -7.0$, $C = 3.5$, $S/\sigma_{lv} = 0.203$. The numerical result is compared to the modified Young equation with the line tension given by Eq. (2.55). Note that the plot is in $1/R$, and hence signals the deviation for smaller droplets. The next-to-leading order $\Delta = \cos\theta - \tau/(\sigma_{lv}R) - \cos\theta_r$ is in $1/R^2$, as seen in the inset in Fig. 2.13.

We close this discussion with remarks on the limit in Eq. (2.55). One finds that the reasoning is only correct provided $m \geq 3$, i.e., in presence of retarded and non-retarded van der Waals forces. For still longer-ranged forces, $2 < m < 3$, the integral is still convergent, but the next-order term behaves according to $R^{-(m-1)}$. For $m \leq 2$, the integral diverges and molecular details then need to be considered.

2.4.3 Line Tension II: The Elasticity of the Contact Line

In this section we look at the (elastic) response of the localized interface to a local distortion. If we try to enlarge a flat liquid interface, we have to pay a price in surface free energy which is proportional to the surface tension. When we disturb the line, we might again expect to run into a line tension. Is this so—and is this object related to the line tension we discussed so far? Again, we follow a discussion presented by Dobbs (1999b). We expect that the *work per unit length* $\delta \mathcal{H}$ can be written as a quadratic function of the amplitudes

$$\delta \mathcal{H}(q) = \frac{1}{2} W(q) \eta_q^2 \tag{2.58}$$

where the η_q are Fourier coefficients. Our interest is in the function $W(q)$.

We address this problem by assuming as the starting point a one-dimensional interface profile in x-direction which is translationally invariant in y-direction. The effective interface Hamiltonian reads as

$$\mathcal{H}[h] = \frac{1}{L_y} \int dx dy \left[\frac{\sigma_{lv}}{2} (\nabla h)^2 + V(h) \right] \tag{2.59}$$

at coexistence, $\Delta \mu = 0$. Without a distortion, the interface profile fulfills the variational equation

$$\frac{\sigma_{lv}}{2} h'(x)^2 = V(h) \tag{2.60}$$

and we assume the boundary condition $h(x) \to h_0$ for $x \to -\infty$. For $x \to \infty$, the profile grows monotonously with the limiting behaviour for $h \to \infty$ as $\sigma_{lv}(\tan \theta)^2 = -2S$, with $S < 0$ at partial wetting.

In order to determine the 'line tension' associated with this configuration we have to introduce the dividing line again, but since we are now looking at an unbounded profile in one direction, we have to modify our expressions accordingly. We use the following form for the harmonic approximation

$$\tau = \sqrt{2\sigma_{lv}} \int_{h_0}^{\infty} dh \left[V(h)^{1/2} - (-S)^{1/2} \right]. \tag{2.61}$$

Distorting the line according to

$$h(x, y) = h(x) + \xi(x, y) \tag{2.62}$$

we can obtain a quadratic expression in ξ:

$$\delta \mathcal{H}^{(2)} \equiv \frac{1}{L_y} \int dx dy \xi(x, y) \left[-\sigma_{lv} \nabla^2 + V''(h(x)) \right] \xi(x, y), \tag{2.63}$$

Fig. 2.14 Local translation
of the interface

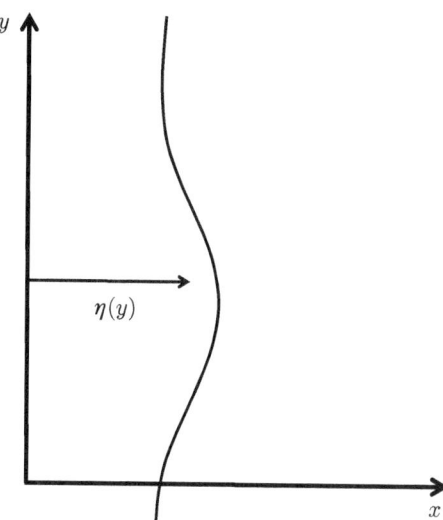

where the effective potential term now contains the whole interfacial profile. The
expression in brackets

$$\mathcal{M} \equiv \left[-\sigma_{lv}\nabla^2 + V''\big(h(x)\big) \right] \tag{2.64}$$

looks like a *Schrödinger operator* of a two-dimensional Hamiltonian. Figure 2.15
displays the interfacial profile, the potential and the ground state wave function to
this operator. An eigenstate of the Hamiltonian is easily obtained, as it is the translational mode

$$\phi_0(x) = h'(x) \tag{2.65}$$

which has the eigenvalue zero. The fact that the ground state has eigenvalue zero
proves the stability of the configuration; $W(q) > 0$ for any wavenumber q.

   *Task: Show explicitly that ϕ_0 is an eigenstate to the operator $\mathcal{M}$ in
Eq. (2.63) with eigenvalue zero.*

   We now enter a subtle point, marking a difference between the case of an interface between two homogeneous phases, like in bulk, or at the prewetting transition
in the case of interfaces. In order to derive $W(q)$ we would then consider local
translations, see Fig. 2.14

$$h(x, y) = h\big(x - \eta(y)\big) \tag{2.66}$$

and we were to obtain

$$W(q) = \frac{\sigma_{lv}}{2}q^2 \int_{-\infty}^{\infty} dx\,\phi_0^2. \tag{2.67}$$

In the present case, this reasoning does not hold because the integral does not con-
verge in the case of a contact line at partial wetting since $\phi_0(x)$ tends to $\sim \tan\theta$
on one side far away from the contact line and therefore the integral diverges. We

Fig. 2.15 The interface
profile $h(x)$, the potential
$d^2V/dh^2(h(x)) \equiv V''(x)$ and
the (non-normalized) 'ground
state wavefunction' $\phi_0(x)$ for
the exemplary interface
potential $V(h) =$
$h^2/(h+1)^4$. For large x,
$h \sim \sqrt{x}$, while
$\phi_0(x) \sim x^{-1/2}$. The potential
$V''(x)$ has a very broad
(undiscernable) maximum at
$x = 6.8$ and decays to zero
for large x from above.
Adapted from Dobbs (1999b)

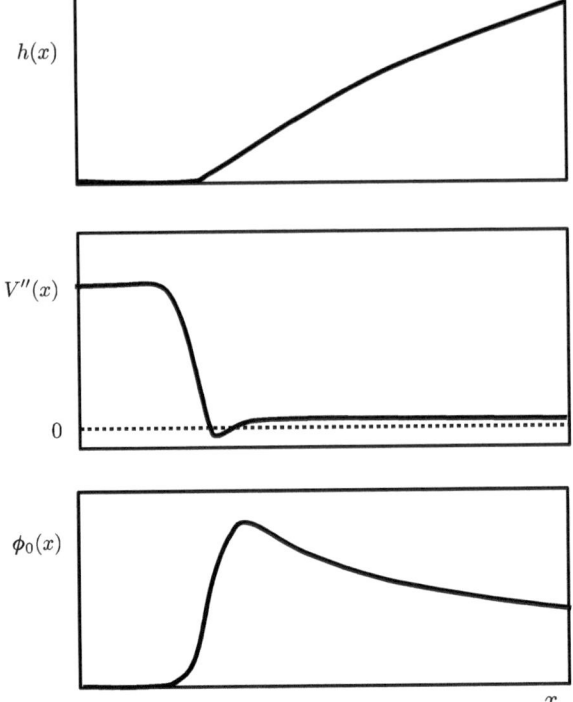

thus have to introduce a constraint in order to render the calculation mathematically
meaningful.

If we call x_c the position of the undistorted line, $h(x_c) = h_c$, the displacement of
the distorted line is given by

$$\eta(y) \equiv -\xi(x_c, y)/\phi_0(x_c) \tag{2.68}$$

to first order in ξ. This corresponds to measuring the displacement with respect to
a particular value of a 'level-set' introduced by h_c. With the choice of a periodic
distortion

$$\xi(x, y) = -\eta_q \phi_q(x) \sin(qy) \tag{2.69}$$

the function $\phi_q(x)$ must then satisfy

$$\phi_q(x_c) = \phi_0(x_c). \tag{2.70}$$

Inserting the expression (2.69) into the expression for $W(q)$ one obtains after integration over y the dispersion relation

$$W(q) = \frac{1}{2} \int_{-\infty}^{\infty} dx \phi_q(x) \left[-\sigma_{lv} \left(\frac{d^2}{dx^2} - q^2 \right) + V''(h(x)) \right] \phi_q(x). \tag{2.71}$$

We now minimize this integral with respect to ϕ_q under the constraint (2.70), which we perform by adding a corresponding term with a Lagrange multiplier λ_q. The resulting Euler-Lagrange equation is

$$\left[-\sigma_{lv}\left(\frac{d^2}{dx^2} - q^2\right) + V''(h(x))\right]\phi_q(x) = \lambda_q\delta(x - x_c) \tag{2.72}$$

with the δ-function $\delta(x - x_c)$. While in the standard case we could approximate ϕ_q with its ground state ϕ_0 to simply have Eq. (2.67), in the present case we have to really determine $\phi_q(x)$. The reason is that the approximation by the ground state fails since in the standard case there is a gap between the ground state and the lowest exited state while in our case there is a continuum of states lying above the ground state.

The states $\phi_q(x)$ can be calculated from a WKB-based matching analysis. Details can be found in Dobbs (1999b); we here only sketch the solution.[4] First, the solution is divided into an 'inner' and an 'outer' region whose matching point x_m is defined by the condition

$$\frac{d}{dx}\left(V''(h(x)/\sigma_{lv})\right) \approx \left(q^2 + V''/\sigma_{lv}\right)^{3/2}. \tag{2.73}$$

In the outer region $x > x_m$ a WKB-approximation yields $\phi_q(x) \sim \exp(-qx)$. In the inner region the potential V'' varies rapidly and the WKB-approximation fails. Instead, one writes

$$\phi_q(x) = \phi_0(x)e^{-S(x)} \tag{2.74}$$

where the function $S(x)$ is continuous at x_c, and $S(x_c) = 0$. The jump in derivative at x_m is proportional to λ_q. $S(x)$ fulfills a second-order differential equation which can be solved in an expansion in wavevectors. One finds

$$S(x) = \begin{cases} \mathcal{O}(q^2), & x < x_c \\ a(q)\int_{x_c}^x d\tilde{x}\phi_0^{-2}(\tilde{x}) & x_c < x < x_m \end{cases} \tag{2.75}$$

which matches the outer solution

$$\phi_q(x) = \phi_0(x_m)e^{-S(x_m)-q(x-x_m)} \tag{2.76}$$

where $a(q) \approx q(\tan\theta)^2$ for partial wetting, and of higher order in q at wetting. In sum, one finds for $W(q)$

$$W(q) = \frac{\sigma_{lv}}{2}(\tan\theta)^2|q| + \frac{\tilde{\tau}}{2}q^2 + \mathcal{O}(|q|^3), \tag{2.77}$$

with

$$\tilde{\tau}(S) = \sqrt{2\sigma_{lv}} \int_{h_0}^{\infty} dh\left[V^{1/2} - \Theta(h - h_c)S^2V^{-3/2}\right], \tag{2.78}$$

---

[4]We will encounter this kind of matching procedure also in Part II of the book in the discussion of dynamic interface profiles.

which in general *differs* from the expression for the line tension, Eq. (2.56). This obviously reflects the nature of the imposed deformation and it also makes clear that in an experimenter has to be careful what quantity is actually looked at.

It is particularly noteworthy that this result depends on the *choice of the position of the contact line*, $h_c = h(x_c)$. This is true at partial wetting, as the result shows, while for $S = 0$, i.e. at wetting, τ and $\tilde{\tau}$ coincide. The approach to the wetting point goes for both expressions with the same power-law dependence in S, but with different amplitudes. The linear $|q|$-dependence is a classic result, see de Gennes (1985).

From the results in this and in the previous section we have learned the first facts about of spatially varying solutions to the (augmented) Young-Laplace equation. In particular, we have clarified the previous purely phenomenological results on the modified Young-equation and learned more about the properties of the line tension at partial wetting, and it also clarifies the interpretation of the experiments on line tensions, as described in Sect. 2.1. There we had also realized that the associated characteristic scales are on the order of 1 nm, hence a truly microscopic length scale. This immediately limits the usefulness and applicability to our interface-model approach, and leaves room for more microscopic approaches, like density-functional theory. We will come to this approach briefly in the next chapter and here only point to recent work by Weijs et al. (2011) describing a study of the line tension combining Molecular Dynamics (MD) simulations and density-functional theory (DFT).

We now turn to a detailed discussion of the *metastable states* that can be identified in the phase diagram of Fig. 2.16, based on the interface potential approach.

2.5 Characterizing Metastable Thin Films

In this section we discuss the properties of one kind of nonequilibrium wetting states which are *metastable states*. Such states are transient states which arise when, briefly stated, the following conditions are met:

- the effective interface potential has a barrier separating a thin from a thick equilibrium state as in Fig. 2.9;
- if $S > 0$ and a thin film is overheated from below to above the prewetting line;
- if $S < 0$ and a thick film is undercooled to below the prewetting line.

In the case $S > 0$, the thin film minimum lies energetically above the thick film minimum, while the opposite is the case for $S < 0$. For $S > 0$, the thin film has thus become a thermodynamically metastable state since it is still locally confined to a parabolic potential, however it has become *globally unstable*. Its confinement to a local parabolic potential, however, does not render it unstable to simple Gaussian fluctuations. A *nonlinear fluctuation* must arise, and in the case of a metastable thin film, this is a droplet of the liquid phase. It must reach a critical free energy (or, simply, *size*) to be able to surmount the free energy barrier between the two states.

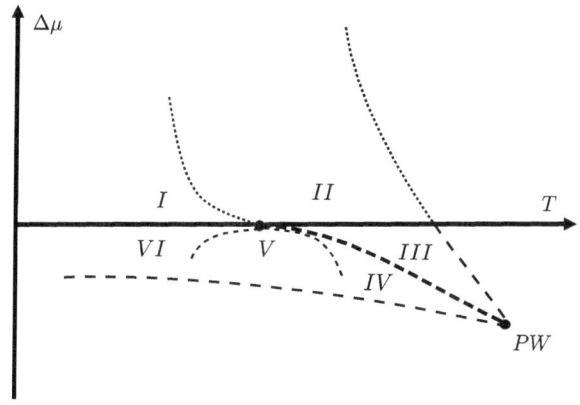

Fig. 2.16 Wetting-dewetting phase diagram in the temperature-chemical potential plane. Indicated are the upper and lower surface spinodals (*thin dashed* and *dotted*), the first-order wetting transition point (*W*), the prewetting line (*thick dashed*) and the prewetting critical point (*PW*). The roman numerals I–VI indicate different scaling regimes of the critical nuclei (droplet and holes) in the phase diagram and are discussed in detail in the text

This critical free energy is, precisely, the *excess free energy* of a *critical droplet*. The critical droplet, and by analogy the *critical hole*, are thus *the essential objects* to characterize a first-order wetting/dewetting transition since the nucleation rate is given by the expression

$$I = D_0 A_0 \exp - \frac{E_c}{k_B T} \qquad (2.79)$$

where D_0 is a prefactor depending on the dynamics of the system, while A_0 is a static prefactor determined by the Gaussian fluctuations of the critical droplet or hole. The inverse of the nucleation rate is the lifetime of the metastable state.

Our starting point for the calculation of the properties of the critical droplet as well as the critical hole is yet again the effective interface Hamiltonian—as we already used it before, but we now only consider the version with a linearized capillary term,

$$\mathcal{H}[h] = \int d^{d-1}x \left[\frac{\sigma_{lv}}{2} (\nabla h)^2 + V(h) - (\Delta \mu)h \right]. \qquad (2.80)$$

This simplification is warranted here since we will be only interested in the *scaling behaviours* of the droplet and hole geometry and the excess free energy in the vicinity of coexistence lines. For the metastable wetting and dewetting states the critical nucleus is again determined by the variational problem

$$\frac{\delta \mathcal{H}[h(r)]}{\delta h(r)} = 0 \qquad (2.81)$$

which leads to the nonlinear elliptic differential equation

$$-\sigma_{lv} \Delta_r h(r) + V'\big(h(r)\big) - \Delta \mu = 0, \qquad (2.82)$$

where Δ_r now is the cylindrical Laplace operator (the radial Laplacian in two dimensions). The critical droplet and critical hole solutions to Eq. (2.82) correspond to two different sets of boundary conditions. We now discuss both types of solutions.

The different possible configurations of critical droplets and critical holes can more precisely be discussed in the context of the wetting phase diagram which allows us to delimit the regions of metastability, see Fig. 2.16. A metastable wetting state can be generated by overheating a microscopically thin layer from $T < T_p(\Delta\mu)$ into a nucleation region bounded by $T_p(\Delta\mu)$ and an *upper spinodal line* $T_{us}(\Delta\mu)$. The spinodal line is determined by the condition $V'(h) - \Delta\mu = V''(h) = 0$. The transition to the stable thick-layer configuration occurs via the random formation of droplets on the thin layer and the growth of supercritical droplets. Close to the prewetting line the critical droplets have a cylindrical shape with a diverging radius at $T_p(\Delta\mu)$. This had already been pointed out early on by Joanny and de Gennes who chose the name 'pancake droplets' for this type of critical nuclei (Brochard-Wyart et al. 1991; Joanny and de Gennes 1984). For critical holes, in total three distinct scaling regimes exist: partial wetting (I), complete wetting (II) and prewetting (III), as we will see.

Dewetting by nucleation of holes can occur when a thick wetting layer from $T > T_p(\Delta\mu)$ is undercooled into a second nucleation region located between $T_p(\Delta\mu)$ and a *lower spinodal line* $T_{ls}(\Delta\mu)$. In this case the critical nuclei are holes in the layer which near $T_p(\Delta\mu)$ are mirror images of the pancake droplets. As we will see, however, close to the partial-wetting line the critical holes have a funnel-like profile with a diverging depth H_c but a finite inner radius R_c at $\Delta\mu = 0$. There exists a third regime, adjacent to the wetting transition point $T = T_w$, $\Delta\mu = 0$, at which H_c and R_c both diverge. We refer to these regimes as the pre-dewetting regime (IV), the complete dewetting regime (V), and the partial-dewetting regime (VI). The three scaling regimes for the critical holes mirror the scaling regimes for the critical droplets.

2.5.1 Wetting by Nucleation: The Critical Droplet

We first discuss the existence of critical droplet solutions, i.e. solutions to Eq. (2.82) which fulfill the boundary conditions (see the sketch in Figs. 2.17, 2.18)

$$h'(0) = 0, \qquad h(\infty) = h_0 \tag{2.83}$$

where h_0 is the thickness of the thin film state, i.e., the microscopic film on the substrate.

An existence proof for a critical droplet solution to Eq. (2.82) is easily sketched by a mechanical analogy based on argument originally made by Sidney Coleman in the context of bulk nucleation (Coleman 1985) and was applied to wetting in Bausch and Blossey (1991). The argument can also be made rigorous (Berestycki

et al. 1981). First we make use, as for the mesoscopic droplet we discussed at the
beginning of the chapter, of the cylindrical symmetry of the droplet and then have

$$-\sigma_{lv}\left(h''(r) + \frac{d-2}{r}h'(r)\right) + V'\big(h(r)\big) - \Delta\mu = 0. \tag{2.84}$$

In the mechanical analogy one interprets the variational equation (2.84) as an equation of motion of a classical particle in time $t \equiv r$ moving in the inverted potential $W(h) = -\Phi(h)$. The potential then has a typical double-well structure for a
finite chemical potential difference, or a very asymmetric shape with one minimum
shifted to $+\infty$ if $\Delta\mu = 0$, i.e. at coexistence of the two bulk phases. Let us first
consider this case.

The solution is obtained if an initial value h_c is found such that for $r \to \infty$
the solution reaches the maximum of $W(h)$ at $h = h_0$. This solution is the unique
case limiting two other types of solutions, whose existence is readily established.
If we start a solution 'far out' from the maximum (large h_c), and still away from
local minimum, the solution will spend a long time in the fairly flat region of the
potential, and still be at a point for which $W(h) > W(h_0)$. Since the time-dependent
friction term $\sim 1/r$ by then has become very small, it can then be neglected and
the remaining potential energy of the solution will lead the particle to overshoot the
maximum. On the other hand, if we start the particle such that $W(h_1) = W(h_0)$, the
friction term suffices to turn the solution into an oscillatory, undershooting solution
which ends up in the minimum of $W(h)$. The physical solution is the limiting one
between the undershooting and overshooting solutions.

In the case $\Delta\mu \neq 0$, the same reasoning holds; only now an overshooting solution
is constructed by placing the starting point for the solution close to the thick-film
maximum. Then again the solution spends a long time in the vicinity of the maximum so that the friction-term can be neglected in its further motion.

Based on the behaviour of these limiting solutions in the different regions of the
wetting phase diagram we can quantify the three scaling regimes for critical droplets
(I–III in Fig. 2.16) as follows.

- *Partial wetting.* If the wall of the system is brought from the gaseous phase to
  the liquid phase (from $\Delta\mu < 0$ to $\Delta\mu > 0$), droplets condense at the wall. This is
  the classic case of *heterogeneous nucleation* at a wall (Blossey 1995). A droplet
  sitting at the wall simply has less volume than a critical droplet in the bulk, and
  hence nucleation at the wall is favoured. In our system, this is the case on the
  liquid side of the bulk phase diagram, $\Delta\mu \to +0$, for temperature $T < T_w$. These
  droplets essentially are a generalization of the droplets discussed at the beginning
  of this chapter. Since we kept the dependence on spatial dimension in our formula
  (2.84)—an approach learnt from the theory of phase transitions—we obtain the
  general results for the height of the critical droplet H_c, its radius R_c and the excess
  free energy E_c (Bausch and Blossey 1993):

$$H_c = (d-1)\frac{|S|}{\Delta\mu}, \qquad R_c = (d-1)\sqrt{2\sigma_{lv}}\frac{|S|^{1/2}}{\Delta\mu} \tag{2.85}$$

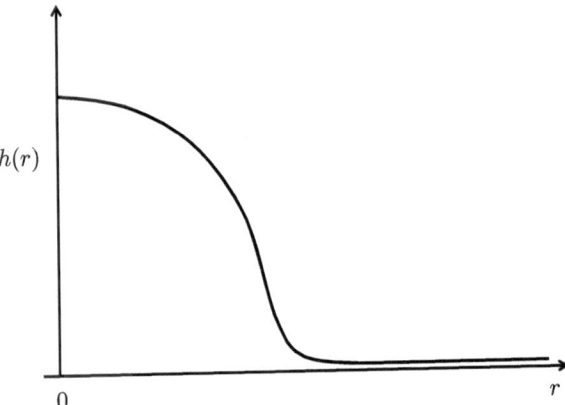

Fig. 2.17 Sketch of the radial profile of a critical droplet

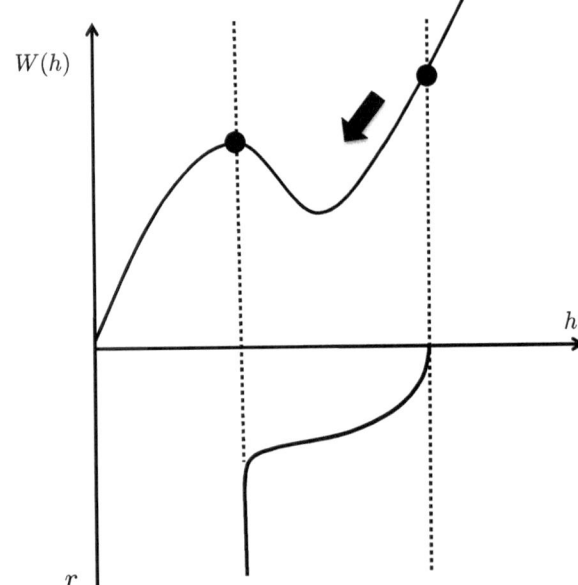

Fig. 2.18 Construction of the droplet solution. For the explanation, see text

and

$$E_c = \Omega_{d-1} 2^{(d+1)/2} \sigma_{lv}^{(d-1)/2} \frac{(d-1)^{d-2}}{d+1} \frac{|S|^{(d+1)/2}}{(\Delta\mu)^{1-d}} \tag{2.86}$$

where Ω_{d-1} is the surface of the $(d-1)$-dimensional unit sphere.

- *Complete wetting.* Here, the droplets are found to scale according to

$$R_c \sim H_c^{(m+1)/2} \tag{2.87}$$

and

$$E_c \sim R_c^{d-2} \ln R_c \tag{2.88}$$

for $m = 3$, non-retarded van der Waals forces, and

$$E_c \sim R_c^{d-2} \tag{2.89}$$

if the forces fall off more rapidly, $m > 3$.

- *Prewetting.* In the vicinity of the prewetting line, the droplets are pancake-like (as stated before) with a scaling behaviour

$$R_c \sim \left[\Delta\mu - \Delta\mu_p(S)\right]^{-1} \tag{2.90}$$

and

$$E_c \sim \left[\Delta\mu - \Delta\mu_p(S)\right]^{2-d} \tag{2.91}$$

where $\Delta\mu_p(S)$ is the prewetting line.

The appearance of logarithmic corrections in the scaling behaviour of critical droplets was noted by Joanny and de Gennes (1984). The paper by Bausch and Blossey (1993) accounts for the full scaling theory of critical droplets at first-order wetting phase transitions, including the crossover between the three scaling regimes, a point we do not pursue here.

The approach also allows the determine the line tension of the drop, which allows to make contact with our previous discussion. At coexistence it scales upon approaching the wetting transition point according to

$$\tau \sim \begin{cases} |S|^{(m-3)/[2(m-1)]} & m < 3 \\ \ln|S| & m = 3 \\ const. & m > 3. \end{cases} \tag{2.92}$$

We see that for $m < 3$ the line tension τ diverges as $S \to 0$, i.e., at complete wetting. This divergence has played a (rather confusing) role in the interpretation of experimental results of dewetting experiments in helium films, see Chap. 3.

This brings us to the properties of critical holes.

2.5.2 Dewetting by Nucleation: The Critical Hole

The link between a bulk-controlled nucleation regime at low temperatures and a surface-dominated nucleation regime above the wetting temperature gives the nucleation phenomenon at wetting already a much richer phenomenology than is the usual case. The specific geometry of the wetting phase diagram has another peculiar feature which shows up in the behaviour of undercooled films. While droplet nucleation is conceptually fairly similar to droplet nucleation in the bulk, hole nucleation is very different (Bausch and Blossey 1994).

The reason for this is very easily understood by a simple consideration of the path of undercooling, and the lines it crosses in the wetting phase diagram, see Fig. 2.19.

Starting point is a stable thick film at an interior point of the prewetting region of the phase diagram close to bulk coexistence, but of course on the gas side. Undercooling this film puts it on a trajectory which crosses the prewetting line and then

Fig. 2.19 A typical
undercooling path of a thick
film to below the prewetting
line

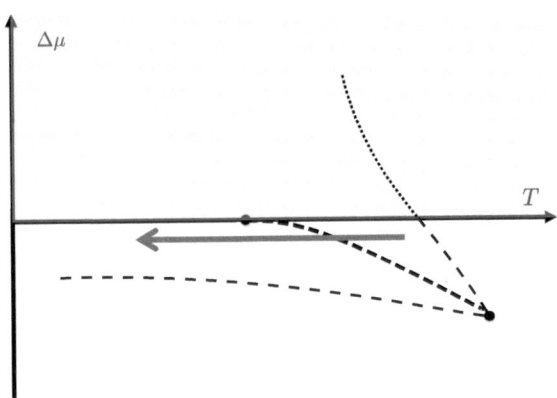

Fig. 2.20 Schematic sketch
of the radial profile of a
critical hole

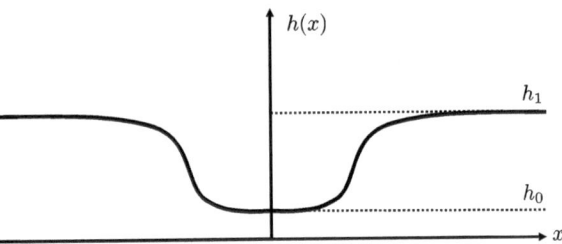

continues on to well below the wetting temperature T_w. The final destination thus
is a point close to bulk coexistence, $\mu \approx \mu_c$. Near bulk coexistence at $T < T_w$, the
size of critical nuclei and their excess free diverge.

The critical holes will have peculiar properties, which we now deduce. The approach follows Foltin et al. (1997). In order to do this, we first specify the two
boundary conditions to equation (2.84). They are, see Fig. 2.20

$$h'(r=0) = 0, \qquad h(r=\infty) = h_1 \qquad (2.93)$$

where h_1 is the film thickness of the flat film at infinity. We note that for the case
at coexistence, $\Delta\mu = 0$, $h_1(\infty) = \infty$ is allowed (and is related to the peculiarity of
the critical hole).

As for the critical droplets, we can argue for the existence of critical holes in the
vicinity of the prewetting line with the shooting argument presented above. Near
interior points of the prewetting line the potential $\Phi(h)$ has a double-well form
which for the radial profile of a critical hole leads to a kink solution. In the limit $T \to$
$T_p(\Delta\mu)$ the position R_c of the turning point of the kink runs to infinity, resembling
the behaviour of a pancake droplet.

However, when some point on the line $\Delta\mu = 0$, $T \leq T_w$ is approached, h_1 and
the critical hole depth $H_c \equiv h_1 - h_0$ diverge. In this regime the critical-hole profile
at macroscopic distances from the wall is determined by the asymptotic behaviour
of $V(h)$ for $h \to \infty$. For long-range molecular interactions this is, as we have seen
before,

$$V(h) = Ah^{1-m}, \quad \text{for } m > 1 \tag{2.94}$$

where A is the (redefined) Hamaker constant, and $m = 3$ or $m = 4$ for non-retarded or retarded van der Waals forces, respectively. By extrapolation the macroscopic profile $H(r)$ of a critical hole can then be defined as the solution of the differential equation

$$\sigma_{lv}\left(H''(r) + \frac{d-2}{r}H'(r)\right) = -A(m-1)H^{-m} - \Delta\mu \tag{2.95}$$

with the new boundary conditions $H(r = R_c) = 0$ and $H(r = \infty) = h_1$ at $\Delta\mu < 0$ or $H'(r = \infty) = 0$ at $\Delta\mu = 0$, respectively. Undershooting and overshooting solutions can still be created, but now by controlling the initial velocity $H'(r = R_c)$.

Since for $r \to R_c$ the friction term and the field $\Delta\mu$ can be neglected in Eq. (2.95), one obtains the analytical result

$$H(r) = \left[\frac{A}{2\sigma_{lv}}(m+1)^2\right]^{\frac{1}{m+1}}(r - R_c)^{\frac{2}{m+1}} \tag{2.96}$$

which is asymptotically valid for all $\Delta\mu \leq 0$.

For $r \to \infty$ and $\Delta\mu \neq 0$ a linear expansion of Eq. (2.95) in $(H_c - H(r))$ leads to a Bessel-type differential equation. This implies an asymptotic form

$$H(r) = H_c\left[1 - C\left(\frac{r}{R^*}\right)^{\frac{2-d}{2}}e^{-r/R^*}\right] \tag{2.97}$$

where $R^* \equiv [Am(m-1)/\sigma_{lv}]^{-1/2}H_c^{(m+1)/2}$, and C is an integration constant. If, as an approximation to the full solution of Eq. (2.95), the expressions (2.96) and (2.97) and their derivatives are matched at some value $r = R_m$, it turns out that $R_m \sim R^*$ and the constant C is of order 1.

For $r \to \infty$ and $\Delta\mu = 0$ the left-hand side in Eq. (2.95) dominates for dimensions $d < d_1(m)$ where the dimension $d_1(m)$ is defined by

$$d_1(m) \equiv \frac{3m+1}{m+1}. \tag{2.98}$$

This leads to the behaviour

$$H(r) = H^*D\left(\frac{r}{R_c}\right)^{3-d} \tag{2.99}$$

for the critical hole profile. Here, the amplitude

$$H^* \equiv \left[A(m+1)^2/8\sigma_{lv}\right]^{1/(m+1)}R_c^{2/(m+1)} \tag{2.100}$$

has been adopted from the exact solution[5]

$$\left(\frac{r}{R_c}\right)^2 - \left(\frac{H}{H^*}\right)^{\frac{m+1}{2}} = 1 \tag{2.101}$$

---

[5]Note that the solution with the '+'-sign corresponds indeed to a droplet solution in the same dimension, $d = d_0(m)$: (Bausch et al. 1994).

of Eq. (2.95) at $\Delta\mu = 0$ in the special dimension

$$d_0(m) \equiv \frac{3m - 1}{m + 1},\tag{2.102}$$

so that $D = 1$ in $d = d_0(m)$.

Due to the boundary condition $H'(r = \infty) = 0$ the asymptotic form (2.99) implies that critical holes at $\Delta\mu = 0$ only exist in dimensions $d > 2$. The necessity of the previously mentioned additional condition $d < d_1(m)$ will become clear through the following analysis in which we map the ordinary differential equation of second order for the droplet profile to a dynamical system.

2.5.3 Mapping to a Dynamical System

Further insight into the solutions can be gained by mapping the ODE to a system of first-order equations, with the peculiarity that, for $\Delta\mu \neq 0$, the system can be represented by three equations. This transformation was conceived by G. Foltin; for the original treatment, which we closely follow, see Foltin et al. (1997).

We define the dimensionless quantities

$$X \equiv \frac{r H'(r)}{H(r)},$$

$$Y \equiv \left(\frac{m^2 - 1}{2}\right) \frac{A}{\sigma_{lv}} \frac{r^2}{H^{m+1}(r)},\tag{2.103}$$

$$Z \equiv -\frac{1}{2} \frac{\Delta\mu}{\sigma_{lv}} \frac{r^2}{H(r)},$$

and consider their dependence on the time-like variable

$$t \equiv \ln\left(\frac{r}{r_1}\right)\tag{2.104}$$

where r_1 is an arbitrary reference scale. Due to (2.4) we find the set of first-order ordinary differential equations in (X, Y, Z),

$$\dot{X} = (3 - d)X - X^2 - \frac{2}{m + 1}Y + 2Z,$$

$$\dot{Y} = 2Y\left(1 - \frac{m + 1}{2}X\right),\tag{2.105}$$

$$\dot{Z} = 2Z\left(1 - \frac{1}{2}X\right).$$

Fig. 2.21 The flow diagram
of the dynamical system
(2.105) with all fixed points
and their principal directions.
The *shaded region* is the
sector in the plane
$X = 2/(m + 1)$ penetrated by
the physical trajectories for
$\Delta\mu < 0$; redrawn after
(Foltin et al. 1997)

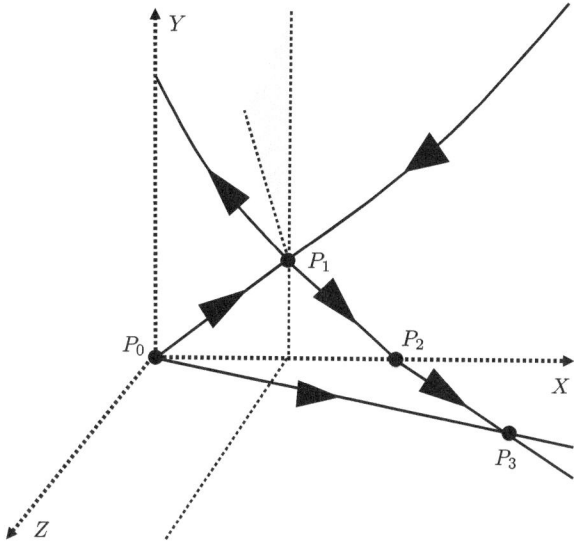

This system has four fixed points

$$X_0 = Y_0 = Z_0 = 0,$$

$$X_1 = \frac{2}{m+1}, \qquad Y_1 = d_1(m) - d, \qquad Z_1 = 0,$$

$$X_2 = 3 - d, \qquad Y_2 = Z_2 = 0, \tag{2.106}$$

$$X_3 = 2, \qquad Y_3 = 0, \qquad Z_3 = d - 1,$$

and we will now discuss their properties.

For $1 < d < d_1(m)$ the fixed points (2.106) are all located in the octant $X \geq 0$,
$Y \geq 0$, $Z \geq 0$, where the critical-hole trajectories occur (whereas $H'(r) \leq 0$ implies
$X \leq 0$ for critical droplets). The subscripts of the fixed-point coordinates indicate
the numbers of the attractive principal directions of each of these points.

In the plane $Z = 0$ the fixed point P_1 in Fig. 2.21 attracts the physical trajectories
coming from $X = Y = \infty$ which then either run to the droplet region $X < 0$, or to
the more stable fixed point P_2. The first possibility corresponds to undershooting
solutions of the saddle-point equation, whereas the second one describes solutions
that obey the boundary conditions for critical holes at $\Delta\mu = 0$. The fixed-point
value $X_2 = 3 - d$ in connection with the definition of X reproduces the asymptotic
behaviour (2.99) up to an undetermined prefactor. In the limit $d \to d_1(m)$ the fixed
point P_1 merges into P_2, and in Fig. 2.21 the right section of the basin of attraction
of P_2 collapses to zero so that critical holes at $\Delta\mu = 0$ no longer exist for $d > d_1(m)$.

If $\Delta\mu < 0$, the physical trajectories approach P_1 from $X = Y = Z = \infty$ but
now have three options to continue. Most of them either run into the droplet region
or to the most stable fixed point P_3, representing undershooting and overshooting
saddle-point solutions where the latter behave as $H(r) = (-\Delta\mu)r^2/[2(d-1)\sigma_{lv}]$

for $r \to \infty$. The basins of attraction for these two sets of trajectories are separated by a surface which is the support of the critical-hole trajectories.

For $d < d_1$ the trajectories for critical holes at $\Delta\mu < 0$ cannot to come close to the fixed point P_2. This is a consequence of the fact that, in agreement with (2.97), these trajectories for $t \to \infty$ have to run to $X = 0$, $Y = Z = \infty$. According to (2.105) they must, however, penetrate the plane $X = 2/(m+1)$ above the line $Z = (2/(m+1))(Y - Y_1)$, and for $X \geq 2/(m+1)$ obey the condition $\dot{Y} \leq 0$. This is incompatible with a visit of the fixed point P_2 which consequently has no influence on the critical-hole profile for $\Delta\mu < 0$.

At bulk coexistence $\Delta\mu = 0$ appears an infinite set of flow lines in the X, Y-plane running from $X = Y = \infty$ to the fixed point P_2. As a consequence the saddle-point equation (2.95) for $\Delta\mu = 0$ has infinitely many solutions which obey the boundary conditions for critical holes. Only one of these solutions will, however, correspond to a true minimum in the variational principle $\delta\mathcal{H}/\delta h = 0$.

The situation can most easily be analyzed in the special dimension $d = d_0(m)$. Here, in terms of the new variables

$$v(t) \equiv \left(\frac{r}{R_c}\right)^{-\frac{2}{m+1}} H(r), \quad t \equiv \ln\left(\frac{r}{R_c}\right), \tag{2.107}$$

the saddle-point equation (2.95) assumes the form

$$\ddot{v} = -\frac{\partial}{\partial v}\left[\frac{A}{\sigma_{lv}}R_c^2 v^{1-m} + \frac{2}{(m+1)^2}v^2\right] \tag{2.108}$$

which again can be considered as an equation of motion for a fictitious particle, but now without a friction term. As a consequence, the particle energy

$$\varepsilon \equiv \frac{1}{2}\dot{v}^2 - \frac{A}{\sigma_{lv}}R_c^2 v^{1-m} - \frac{2}{(m+1)^2}v^2 \tag{2.109}$$

is conserved. The last equation can be rewritten as

$$(x-1)^2 = \frac{2}{m-1}y - \lambda\frac{m+1}{m-1}y^{\frac{2}{m+1}} + 1 \tag{2.110}$$

where

$$x \equiv \frac{X}{X_1}, \qquad y \equiv \frac{Y}{Y_1}, \qquad \lambda \equiv -\frac{\varepsilon}{\varepsilon_0}, \tag{2.111}$$

and $\varepsilon_0 \equiv [2/(m^2-1)][(m-1)(m+1)^2 A R_c^2/(4\sigma_{lv})]^{2/(m+1)}$ is the maximum value of the potential energy in (2.109). With λ taken as a parameter, Eq. (2.110) analytically describes the full flow pattern of the system which is depicted in Fig. 2.22.

Obviously this pattern is symmetric with respect to the line $x = 1$. For $x = y = 1$ one obtains the parameter value $\lambda = 1$ which consequently belongs to the separatrix running through the fixed point P_1 (see Fig. 2.22). This corresponds to zero kinetic energy of the fictitious particle when it passes the maximum of the potential in (2.109). Values $\lambda > 1$ accordingly belong to undershooting solutions, whereas for $\lambda < 1$ one finds an infinite set of solutions obeying the critical-hole boundary

Fig. 2.22 The flow pattern corresponding to Eq. (2.110) for the limiting case $m = 3$. (Note that $H'(r = \infty) = 0$ for $m = 3$) redrawn after Foltin et al. (1997)

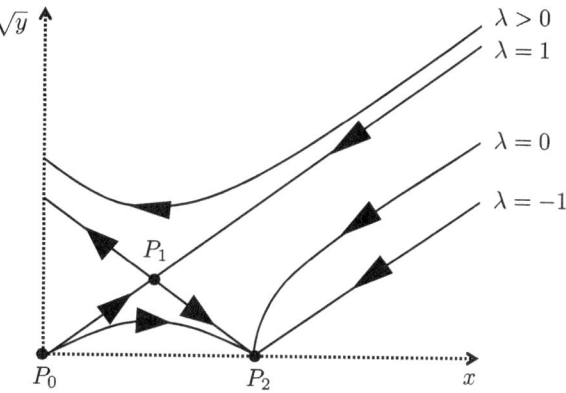

conditions. The profile (2.101) leads to the value $\lambda = 0$, i.e., a simple parabola for the corresponding flow line.

For $d = d_0(m)$ the energy functional $\mathcal{H}[H(r)]$ can be written in the form

$$\mathcal{H} = \Omega_{d-1}\big[\big(m^2 - 1\big)AY_1^m/(2\sigma_{lv})\big]^{2/(m+1)} u[y] \tag{2.112}$$

where

$$u = \int_0^\infty dy \frac{y^{-\frac{m+3}{m+1}}}{2(x-1)}\left[\frac{1}{2}x^2 + \frac{1}{m-1}y\right]. \tag{2.113}$$

This implies

$$\frac{\delta u}{\delta x(y)} = \frac{1}{4}y^{-\frac{m+3}{m+1}}\frac{1}{(x-1)^2}\left[x^2 - 2x - \frac{2}{m-1}y\right], \tag{2.114}$$

which shows that the variational principle $\delta u = 0$ leads to (2.110) with $\lambda = 0$. The special solution (2.101) therefore, in fact, corresponds to a true minimum of $\mathcal{H}$.

2.5.4 Scaling Behaviour of the Critical Hole

We now turn to the characterization of the critical hole in terms of scaling quantities. These are the quantities H_c, R_c, and E_c for a critical hole at arbitrary values of $\Delta\mu$. For this we use the definitions

$$\Phi'(H_c + h_0) = 0, \qquad E_c \equiv \mathcal{H}\big[h(r)\big] - \mathcal{H}[H_c + h_0], \tag{2.115}$$

and extract R_c from the relation

$$\Phi(H_c + h_0) - \Phi(h_0) = (d - 2)\sigma_{lv}\int_0^\infty dr\, r^{-1}\big(h'(r)\big)^2, \tag{2.116}$$

implied by the saddle-point equation. The second Eq. (2.116) can slightly be simplified by use of the virial theorem which states that the potential-energy part in (2.115) is $(3 - d)/(d - 1)$ times the kinetic part. This follows from the scaling property

$$\partial\mathcal{H}\big[h(\alpha r)\big]/\partial\alpha|_{\alpha=1} = 0 \tag{2.117}$$

which in turn is implied by the variational principle $\delta\mathcal{H}[h_\alpha(r)] = 0$ for the special set of functions $h_\alpha(r) \equiv h(\alpha r)$. As a result we obtain

$$E_c = (d-1)^{-1}\Omega_{d-1}\sigma_{lv}\int_0^\infty dr\, r^{d-2}(h')^2 \qquad (2.118)$$

where Ω_{d-1} again is the volume of the $(d-1)$-dimensional unit sphere.

Close to the line $\Delta\mu = 0$, $S \le 0$, we can in (2.115)–(2.118) neglect the microscopic increment h_0 to H_c, replace $V(h)$ by its asymptotic form and insert for $h(r)$ the macroscopic profile $H(r)$. This leads to the result

$$H_c = \frac{1}{(m-1)A}|\Delta\mu|^{-\frac{1}{m}} \qquad (2.119)$$

for $\Delta\mu \to 0$, and, to leading order, to the equations

$$AH_c^{1-m} - \Delta\mu H_c - S = (d-2)\sigma_{lv}\int_{R_c+r_0}^\infty dr\, r^{-1}(H')^2(r), \qquad (2.120)$$

$$E_c = \frac{\sigma_{lv}\Omega_{d-1}}{(d-1)}\int_{R_c+r_0}^\infty dr\, r^{d-2}(H')^2(r). \qquad (2.121)$$

Here, a cut-off length $r_0 \ll R_c$ has been introduced in order to remove the artificial singularity which, due to the extrapolation of $V(h)$ down to small h, occurs at $r = R_c$ in the case $m \ge 3$. In (2.120) and (2.121) we now split off integrals running from $R_c + r_0$ to $(1+\lambda)R_c$ where the choice $0 < \lambda \ll 1$ allows to use Eq. (2.94). In the remaining integrals we transform to the scaled variables r/R^* for $\Delta\mu < 0$ and r/R_c for $\Delta\mu = 0$, suggested by the asymptotic forms of the hole profiles. This leads to a power in R^* and R_c, respectively, where the cofactors are assumed to be finite in the limit $h \to 0$ and $S \to 0$.

On a path $S = const.$ in the partial-dewetting regime the procedure just described leads for $|\Delta\mu| \to 0$ to a constant value of R_c, and to

$$E_c \sim H_c^{\frac{m+1}{2}(d-d_0(m))} \qquad (2.122)$$

with H_c given by (5.4). At $S = 0$, i.e. in the complete-dewetting regime, we find the behaviour

$$\begin{array}{ll} R_c \sim H_c^{\frac{m+1}{2}} & \text{for } m < 3, \\[6pt] R_c \sim H_c^2 \ln H_c & \text{for } m = 3, \\[6pt] R_c \sim H_c^{m-1} & \text{for } m > 3 \end{array} \qquad (2.123)$$

where again H_c has the form (5.4). Moreover, in this regime we obtain

$$\begin{array}{ll} E_c \sim R_c^{d-d_0(m)} & \text{for } m < 3, \\[6pt] E_c \sim R_c^{d-2}\ln R_c & \text{for } m = 3, \\[6pt] E_c \sim R_c^{d-2} & \text{for } m > 3. \end{array} \qquad (2.124)$$

For critical holes at $\Delta\mu = 0$ the only nontrivial result is the behaviour

$$R_c \sim |S|^{-\frac{m+1}{2(m-1)}} \quad \text{for } m < 3,$$
$$R_c \sim |S|^{-1} \ln |S| \quad \text{for } m = 3, \tag{2.125}$$
$$R_c \sim |S|^{-1} \quad \text{for } m > 3$$

for $|S| \to 0$.

In the pre-dewetting regime the asymptotic behaviour of the pancake critical holes will, with growing distance from T_w, increasingly depend on the microscopic details of the potential $\phi(h)$. We therefore have to go back to the relations (2.116) and (2.118), in which we then use the fact that $h'(r)$ is sharply peaked at the value $r = R_c$. This leads for any path $S = const.$ to a constant value of F_c at the prewetting line $\Delta\mu = \Delta\mu_p(T)$, and to the relations

$$R_c \sim \left(\Delta\mu_p(T) - \Delta\mu\right)^{-1}, \tag{2.126}$$

$$E_c \sim \left(\Delta\mu_p(T) - \Delta\mu\right)^{2-d} \tag{2.127}$$

which are identical to those for pancake droplets. When the wetting transition point is approached along the prewetting line, H_c diverges as in (2.119).

The crossover lines, separating the pre-dewetting and the partial-dewetting regime from the intervening complete-dewetting regime (see Fig. 2.16), are of the form $|\Delta\mu| \sim |S|^{\frac{m}{m-1}}$. This is implied by Eq. (2.120) through which the spreading coefficient S enters the calculation in the form $S + const.|\Delta\mu|^{\frac{m-1}{m}}$.

We have achieved a complete description of the (scaling) properties of critical holes. Of course, in case a complete knowledge of system parameters can be achieved, it is possible to explicitly calculate the excess free energy of the critical holes and the nucleation rate, Eq. (2.79).

> *Task. Calculate the Gaussian fluctuation spectrum of a critical hole, i.e., the second variation of the interface Hamiltonian $\mathcal{H}$ at the hole profile. What can be said qualitatively about the lowest-lying modes? See our discussion of the elastic properties of the contact line.*

2.5.5 Undercooling a Thick Film: A Physical Interpretation, and Some Theory

In this section we will discuss the undercooling of a thick film in somewhat more detail. The peculiarity of the undercooling of thick films has indeed been encountered experimentally in both classical liquids and in helium films, we will discuss these results in the following chapter. The interpretation of the experiments has, however, suffered from some conceptual confusions which make it useful to address this point specifically before going to a description of the experiments.

Figure 2.19 shows an undercooling path that is typically encountered experimentally. The system is initially prepared in a thick film state, controlled by its distance from the coexistence line, $\Delta\mu \to -0$, on the gas side of the wetting phase diagram. The film in then undercooled to a point below the prewetting line. Here, one encounters a 'strange' situation, since the critical hole is almost infinite in size and excess free energy for $\Delta\mu = -0$, and strictly infinite at $\Delta\mu = 0$. Thermal nucleation is therefore strongly suppressed.

In view of our previous discussion, there is *nothing* strange with this: the bulk coexistence line is a natural continuation of the prewetting line, except for the fact that one of phases now is infinitely thick (gas = thickness 0, liquid = thickness ∞). All along each coexistence line, the critical nucleus has (by definition) an infinite size and excess free energy. In this picture, the decay of the thick films is thermodynamically disfavoured since the undercooling path leads close parallel to the coexistence line. The reason for the observed large stability of an undercooled film is thus nothing *but the asymmetric topology of the wetting phase diagram.*

Two alternative discussions exist in the literature, and we briefly address why they fail. Schick and Taborek (1992) related the long lifetime of undercooled films to the divergence of the line tension at wetting. It is, however, hard to see how the line tension at wetting can influence a process which never encounters the wetting point, and hence this divergence. The singularity that is encountered comes from a critical hole of almost infinite size. Put in another way: the physics of the problem is unrelated to the line tension, since the topology of the phase diagram does not depend on it. For both diverging and finite line tensions, the undercooling problem is the same.

A second discussion is by Herminghaus and Brochard (2006), who mix two things; the problem of the asymmetry of the interface configuration—a point we addressed in the discussion of the contact line—and the presence of gravity, both of which are claimed to make our discussion obsolete. However, for any finite value of $\Delta\mu < 0$ the critical hole exists and has a finite (but very large) excess free energy. No critical hole of finite excess free energy exists on *any* of the coexistence lines. And, finally, gravity does not play a role in this at all: when films become gravity-controlled, there is no wetting phase transition anymore, see Blossey and Oligschleger (1999). Our whole discussion only makes sense if gravitational effects are not affecting the systems which are solely controlled by dispersion forces.

To round-up this pre-discussion of experiment vs theory, it should be noted that the role of thermal nucleation of a hole in the film and the enormous stability of undercooled wetting layers is strongly affected by the presence of borders—remember the discussion of the dewetting binary fluid mixture in the Introduction. When the system is prepared such that a border is present from which either a thin or a thick film can invade the surface, this process is generally much faster than the nucleation of holes (or, as in the complex dewetting scenario in the Introduction, both processes actually compete).

2.6 Dewetting for an Unstable Film: Spinodal Decomposition

After this very detailed discussion of the properties of a metastable thick film state we can now attack the second, and in fact, at first sight formally simpler case, that of an unstable thick film. In order to discuss this case we return to the interface Hamiltonian (2.2) taken at coexistence, $\Delta\mu = 0$, and at $d = 3$:

$$\mathcal{H}[h] = \int d^2x \left[\frac{\sigma_{lv}}{2}(\nabla h)^2 + V(h)\right],\tag{2.128}$$

and calculate the variation $\delta\mathcal{H}/\delta h(x) = 0$ for a one-dimensional configuration, assuming translational invariance in the transverse direction, which yields

$$-\sigma_{lv}h''(x) + V'(h) = 0.\tag{2.129}$$

We are now interested in a perturbation of a homogeneous film of thickness h_0, at sufficiently large h such that only the asymptotic form of $V(h)$ matters. We then perturb this film via

$$h(x,t) = h_0 + \varepsilon \exp(iqx)\tag{2.130}$$

and then obtain

$$M(h_0) \equiv -\sigma_{lv}h''(x) + V''(h_0)h(x) = \left(-\sigma_{lv}q^2 + V''(h_0)\right)\exp(iqx)$$
$$= \sigma_{lv}\left(q_s^2 - q^2\right)\exp(iqx).\tag{2.131}$$

We can find a wavelength $\lambda_s \equiv 2\pi/q_s$

$$\lambda_s = (2\pi)\sqrt{\frac{\sigma_{lv}}{-V''(h_0)}},\tag{2.132}$$

such that $M(h_0) > 0$ for $\lambda > \lambda_s$ and $M(h_0) < 0$ for $\lambda < \lambda_s$. The wavelength thus defines a change of stability for the perturbation. The wavelength λ_s depends on the curvature of the interface potential; if for the latter we have $A < 0$ in the asymptotic form, λ_s will be meaningful, which is indeed the case for an unstable film.

The wavelength we calculated turns out to be essentially the wavelength of *spinodal dewetting*. (The prefactor is not correct since we so far neglected the dynamics of the film which respects mass conservation; the real wavelength thus turns out to be larger by a factor of $\sqrt{2}$.) For wavelengths larger than λ_s, perturbations of the film will grow, while perturbations of shorter wavelength will relax. For the case of a van der Waals interface potential, $\lambda_s \sim h_0^2$.

The spinodal dewetting scenario has been postulated several times in the literature; the first appears to be by Vrij (1966). In the context of wetting phenomena, the most cited reference is Brochard-Wyart and Daillant (1989).

Task. The previous calculation shows that for each film height h_0 there is a characteristic wavelength λ_s such that for $\lambda > \lambda_s$ the film is unstable to fluctuations. This calculation, however, ignores flow and mass conservation in the film. Assuming a Poiseuille flow one has

$$j = C\frac{d}{dx}\mathcal{H}[h(x)],\tag{2.133}$$

where $C = h^3/3\eta$, where η is viscosity. Calculate the correct spinodal dewetting wavelength if in addition mass conservation in the film is assumed.

In order to close this chapter, we finally address the question of the range of validity of the effective interface approach, at least from a theoretical perspective. For this, we briefly address a more microscopic approach, *density functional theory*.

2.7 Density Functional Theory

The effective interface approach has allowed us to venture beyond the macroscopic concept of Young's equation to include *microscopic effects* on a *mesoscopic scale*: up to several tens of nanometers. Getting down into the nanometer range, our approach will certainly be pushed to its limits, if not beyond. Can we extend our concepts into that range? For more microscopic aspects, the use of *density functional theory* is a possibility. A basic review article on the method is by Evans (1979).

The basis of this approach is the *grand canonical free energy functional* of the number density $\varrho(\mathbf{r})$ of a one-component fluid (Bauer and Dietrich 1999)

$$\Omega[\varrho(\mathbf{r}); T, \mu] = \int_\Lambda d^3 r f_{HS}(\varrho(\mathbf{r}), T) + \int_\Lambda d^3 r [V(\mathbf{r}) - \mu]\varrho(\mathbf{r})$$
$$+ \frac{1}{2} \int_\Lambda d^3 r \int_\Lambda d^3 r' \varrho(\mathbf{r}) w(|\mathbf{r} - \mathbf{r}'|)\varrho(\mathbf{r}') \qquad (2.134)$$

where Λ is a finite fluid volume in the half space Λ_+ with $z > 0$, with $\Lambda \to \Lambda_+$ in the thermodynamic limit. The external potential $V(\mathbf{r})$ describes the interaction of a fluid particle with the substrate via

$$V(\mathbf{r}) = V(z > 0) = -\sum_{i \geq 3} \frac{u_i}{z''}. \qquad (2.135)$$

An exemplary case would be a *Lennard-Jones potential*

$$\phi_v = 4\varepsilon_v \left[\left(\frac{\sigma_v}{r}\right)^{12} - \left(\frac{\sigma_v}{r}\right)^6 \right] \qquad (2.136)$$

where ε_v is the interaction strength and σ_v characterizes molecular size. We take with $v = w$ and assume a lateral average for the wall-fluid interaction V, and with $v = f$ for the attractive part of the pair potential between the fluid particles. Approximately, one has for w the expression

$$w(r) = \int d^3 r w(r) = \frac{4 w_0 \sigma_f^3}{\pi^2} \frac{1}{(r^2 + \sigma_f^2)^3} \qquad (2.137)$$

where $w_0 \sim \varepsilon_f \sigma_f^3$. The repulsive part of the interaction is modeled by the first term in Eq. (2.134) which is a reference free energy of a hard-sphere fluid in a *Carnahan-Starling approximation*

$$f_{HS}(\varrho, T) = k_B T \varrho \left(\ln(\varrho \lambda^3) - 1 + \frac{4\eta_p - 3\eta_p^2}{(1 - \eta_p)^2} \right) \qquad (2.138)$$

where λ is the thermal de Broglie wavelength λ, $\eta_p = (\pi/6)\varrho d^3(T)$ the dimensionless packing fraction, and $d(T)$ an effective hard-sphere diameter.

From this theory the bulk properties are easily derived by minimizing Ω with respect to density, ignoring the wall-fluid interaction. The analysis of film properties then proceeds by introducing a '*sharp-kink approximation*' in the form

$$\widehat{\varrho}(x, z) = \Theta(z - d_w)\{\varrho_l \Theta(h(x) - z) + \varrho_g \Theta(z - h(x))\} \qquad (2.139)$$

where d_w is the location of the wall, and where $h(x)$ asymptotically for large arguments approaches

$$\widehat{h}(x) = h_0 \Theta(-x) + (h_0 + x \tan \theta)\Theta(x), \qquad (2.140)$$

with the contact angle θ; this corresponds to the situation we discussed already for the line tension.

From here on one can separate the contributions to the grand canonical free energy into bulk, surface and line contributions; we refer to Bauer and Dietrich (1999) and the references listed there for details. For our present purposes we state only the Euler-Lagrange equation for the film which is a *nonlocal* expression, due to the presence of the double integral over fluid space:

$$-\Delta\varrho \int_{-\infty}^{\infty} dx' \int_0^{h(x')-h(x)} dz' \overline{w}\big(|x - x'|, |z'|\big) = \big[V\big(h(x)\big) - \varrho_{lt}\big(h(x) - d_w\big)\big] \qquad (2.141)$$

wherein $w(x, z)$ and $t(z)$ are defined as

$$\overline{w}(x, z) = \int_{-\infty}^{\infty} dy\, w\big(|\mathbf{r}|\big) \qquad (2.142)$$

and

$$t(z) = \int_z^{\infty} dz' \int_{R^2} dx\, dy\, w\big(|\mathbf{r}|\big), \qquad (2.143)$$

and ϱ_l is the liquid density, $\Delta\varrho$ the density difference between the two bulk fluid phases (liquid/gas).

How does this nonlocal theory deviate from the local theory we have used? In fact, they differ not by much in the context of the sharp-kink description which underlies both the local theory and the present DFT treatment (Bauer and Dietrich 1999). Against earlier claims based on errors in the numerical treatment of Eq. (2.141), the nonlocal theory reproduces the results of the local theory for the line tension. The only noticeable differences arise with respect to interfacial curvatures. These findings therefore vindicate our treatment—in the limit of the desired precision.

However, there are real corrections to the capillary dispersion theory that can indeed only be captured by a careful treatment of effects starting from the DFT. One can derive an effective interface Hamiltonian which correctly treats long-ranged effects, and then leads to a *wavevector-dependent surface tension*. We refrain from writing down the expression and refer the interested reader to the original derivation (Mecke and Dietrich 1999). Although this is indeed a subtle effect, it has been observed experimentally in water Fradin et al. (2000).

Chapter 3
From Classical Liquids to Polymers

In this chapter we illustrate the previous theoretical discussion with experimental results—this will serve us as a *reality check* (but we have already seen some experimental findings that support our approach in Chap. 2).

The study of the wetting and dewetting phase transitions has seen a major progress in the 1990s when, in parallel, several different experimental systems became available which allowed a direct confrontation between theory and experiment. From our conceptual point of view here, we base the discussion on effective interface models with an interface Hamiltonian of the general form, as we argued before

$$\mathcal{H} = \mathcal{H}_{capillary} + \mathcal{H}_{dispersion} \tag{3.1}$$

and we will show that the different experimental systems provide us with different realizations for $\mathcal{H}$.

3.1 Classical Liquids & Soap Films

Classical liquid mixtures have been among the first systems for which the original wetting scenario by J. Cahn was investigated. The prime example is the *binary liquid mixture* cyclohexane-methanol, for which wetting and prewetting transitions were first established experimentally (Bonn et al. 1992; Kellay et al. 1993). This mixture can be described with our standard interface Hamiltonian with a capillary-wave term and a standard interface potential governed by van der Waals forces. The experimental difficulty encountered in this mixture has been that the prewetting line is very close to bulk coexistence, which made it difficult to map out the phase diagram for a long time. Other classical liquid mixtures that have received interest are binary mixtures of methanol and *n*-alkanes ($C_n H_{2n+2}$). By varying the *n*-alkane length, different types of wetting phase diagrams become accessible. Most interest in these systems was raised due to the fact that *short-range critical wetting* behaviour could

R. Blossey, *Thin Liquid Films*, Theoretical and Mathematical Physics,
DOI 10.1007/978-94-007-4455-4_3, © Springer Science+Business Media Dordrecht 2012

Fig. 3.1 Disjoining pressure
$\Pi(h)$ for a soap film of
1.36×10^{-4} M $C_{12}E_6$ in
10^{-3} M salt NaCl. *Triangles*
refer to pressure increase,
circles to pressure decrease.
Reprinted with permission
from Casteletto et al. (2003).
Copyright by the American
Physical Society

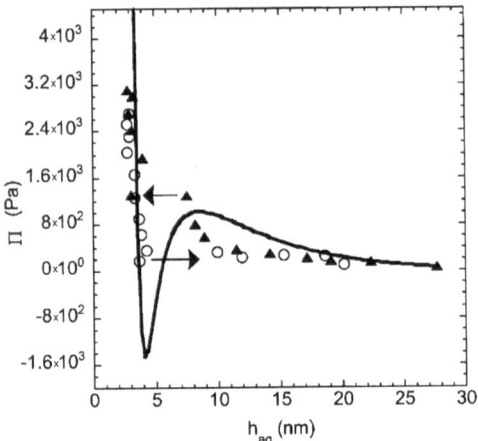

be observed in this system (Ross et al. 1999). Hallmark of short-range critical wetting systems is that have interface potentials of exponential decay,

$$V(h) \sim \exp(-h/\xi) \tag{3.2}$$

where ξ is the correlation length in the fluid which diverges at the critical point, but
is finite at all wetting transition points. For a detailed discussion of these transitions,
see the recent review by Bonn et al. (2009).

   Soap films of nonionic surfactants are, under variation of external pressure, either in a common black film or in a microscopic black film state, a *Newton black
film* state (Bergeron et al. 1992). The transition between the two is hysteretic and
of wetting-dewetting type, and has indeed been discussed already in the 1960s in
this context (De Feijter and Vrij 1972; Vrij 1966). Figure 3.1 shows the experimentally measured disjoining pressure which mathematically is given by the equilibrium
condition

$$\Pi(h) \equiv V'(h) = Ce^{-\kappa_D h} - \frac{A}{h^3} + \frac{B}{h^9} = p_{ext} - p_{atm} = \Delta p. \tag{3.3}$$

The first term stems from electrostatic interactions in the film, with κ_D^{-1} as the
Debye screening length. The second term is the long-range dispersion force with
Hamaker constant A, while the third term is the steric repulsion between the surfac-
tant molecules.

3.2 Liquid Helium

Liquid helium was originally an unlikely candidate for wetting and dewetting phenomena since it generally is the case that a helium molecule interacts more strongly
with all other molecules than with another helium molecule. Unless the other
molecule behaves in a way that helium does not like at all, and this is precisely the

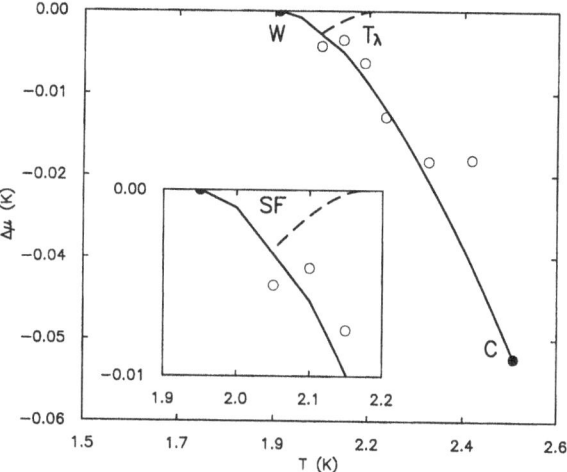

Fig. 3.2 Prewetting phase diagram for ^{4}He on Cs. W is the wetting transition; note the superfluid region in its vicinity, indicated by the λ-line. Reprinted with permission from Cheng et al. (1993). Copyright by the American Physical Society

case for the highly reactive alkali metals like rubidium (Rb) and cesium (Cs). These metals have work functions characteristic of weakly-bound electrons, and since a helium atom has a completely filled shell, it is repelled by those metals. At very low temperatures, helium can thus be brought into a dewetting transition (Nacher and Dupont-Roc 1992; Rutledge and Taborek 1992; Taborek and Rutledge 1992).

The asymmetry of the first-order transitions of wetting and dewetting (see Fig. 3.2), leads, as discussed before, indeed to the very different lifetimes of metastable wetting and dewetting states in the vicinity of the wetting phase transition. This difference was first erroneously ascribed to the divergence of the line tension Schick and Taborek (1992), see the discussion in the previous chapter. As argued there, the experimental path crosses the wetting transition to lower temperatures so that this singularity is, in fact, never encountered.

Liquid helium experiments on the alkali metals are strongly influenced by substrate roughness and disorder. Cs, and even more strongly Rb, cannot easily be handled due to their reactivity and therefore present substantial disorder effects which also have been addressed theoretically. E.g., in the experimental setup by Nacher and Dupont-Roc, the metal film is present as a ring in a capillary tube, creating two compartments in the rings when it is not wetted by He, thus affecting thermal conduction through the tube. In this setup, wetting is brought about by an invasion of the film, not by nucleation of droplets (which is the slower process). Dewetting then proceeds by hole nucleation, but is affected by the 'rough landscape' underlying the film surface.

So far, the quantum nature of liquid helium has not been brought into the picture. This can, e.g., be done by subjecting the system to rotation in order to create vortices. Vortices then can interact with hole formation since their presence affects the free film surface by creating surface dimples: these can be used as hole precursors. This topic is a curious and little noticed aspect which has been theoretically discussed by Blossey (1998) and de Gennes (1999). The effect can even be magnified by adding electrons (Blossey 2001a; Montevecchi and Blossey 2000). None of these aspects has yet been studied experimentally.

3.3 Wetting in Superconductors

Wetting phenomena in superconductors seem to be a strange thing at first sight. It should, however, be kept in mind that (i) surface superconductivity is a classic effect in superconductors, and (ii) that superconductors are, in effect, based on normal and superconducting *electron liquids*. The notion of the interfacial tension $\sigma_{SC/N}$ between the normal- and superconducting phases is well-defined (and a textbook problem) for type-I superconductors.

Wetting (and dewetting) can arise when a type-I superconductor is modified in order to enhance superconductivity near its surface. This can be achieved already by mechanical work done to the surface. Microscopically, the effect can e.g. be due to defects in the crystal lattice.

The wetting-scenario in type-I superconductors was first proposed in Indekeu and van Leeuwen (1995) on the basis of the Ginzburg-Landau theory of superconductors with a modified surface term, and the full phenomenology was developed subsequently. Experimental confirmation was achieved in Kozhevnikov et al. (2007).

From the point of view of interfacial Hamiltonians, the theory is very interesting since both the capillary and the dispersion distributions to $\mathcal{H}$ are highly unusual. The Ginzburg-Landau theory of superconductors contains only short-range interactions, hence it is no surprise that the effective interface potential for large film thicknesses decays exponentially,

$$V(h) \sim \exp(-h/\xi_0). \tag{3.4}$$

Here, ξ_0 is the coherence length of the superconductor. In contrast to classical liquids, it is a huge quantity. The decay of the potential is thus very slow.

For a type-I superconductor, the interface potential can be calculated exactly in the limit $\kappa \to 0$, which is the ratio of magnetic field penetration length to coherence length. The result is shown in Fig. 3.3, as well as the corresponding theoretically predicted wetting phase diagram for this case.

P.G. de Gennes speculated that the difference between the dielectric properties of the normal and superconducting phases could also give rise to long-range forces. Calculations of the effective interface potential from Ginzburg-Landau theory and from the DLP-theory of dispersion forces allow to determine such contributions, with the result (Blossey 2001b)

$$V(h) \sim -\frac{A}{h^5}, \tag{3.5}$$

yielding the interesting situation that the long-range forces are actually fairly short-ranged, while the short-range forces are fairly long-range, see above.

Finally, also the capillary Hamiltonian of a normal-superconducting interface is quite different from that of a liquid; the reason is the presence of the magnetic

Fig. 3.3 *Top*: Wetting phase diagram for a type-I superconductor in the limit $\kappa \to 0$ in the parameters applied magnetic field **H** and temperature. The line FN (for first-order nucleation) corresponds to the prewetting line. *Bottom*: effective interface potential. It is to be noted that the potential starts linearly rather than parabolically near the wall. Reprinted with permission from Blossey and Indekeu (1996). Copyright by the American Physical Society

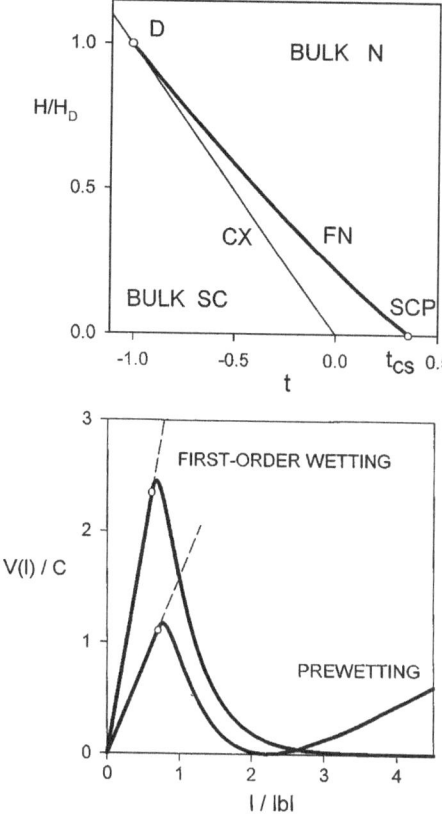

field which renders interfacial fluctuations different in longitudinal and transverse directions (Dobbs and Blossey 2000). In q-space, the Hamiltonian reads as

$$\mathcal{H}_{cap,SC} = \frac{1}{2} \int_{\mathbf{q}<\Lambda} \frac{d^2\mathbf{q}}{(2\pi)^2} \left[\sigma_{SC/N} \mathbf{q}^2 + (\mathbf{q} \cdot \mathbf{H})^2 |\mathbf{q}|^{-1} + m^2 \right] |h_q|^2 \qquad (3.6)$$

The presence of the first term which is effectively $\sim |\mathbf{q}|$. The presence of this term has the consequence that upon the wetting transition, when the interface *unbinds*, it does not simultaneously roughen as a liquid interface would; this follows from a study of the correlation function in the above-mentioned paper.

3.4 Dewetting of Polymer Thin Films

Polymer thin films have proven to be an extremely interesting test case for dewetting studies, for many reasons, among which are the ease of preparation and control of the samples, ease and precision of measurements, practical relevance, complexity of the systems... we will get a full picture of these advantages in the following.

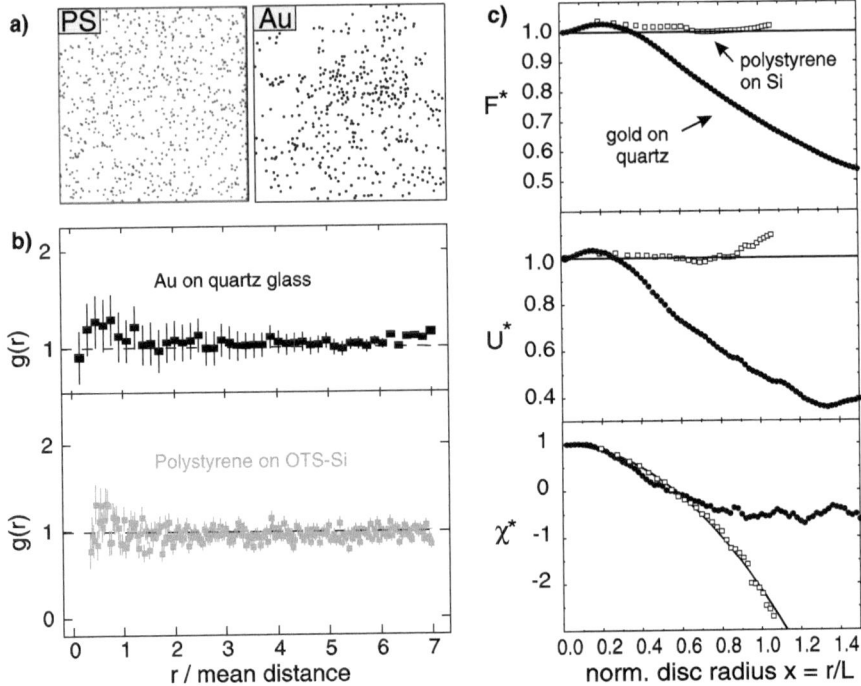

Fig. 3.4 (a) Position of holes in a PS film (*left*) and in a gold film (*right*). (b) The pair correlation function $g(r)$ for the patterns in (a). (c) Normalized Minkowski measures F^*, U^* and χ^* for both substrates (see Appendix B for a brief introduction). The *solid line* which is followed by PS has been generated by a Poisson point process—hence heterogeneous nucleation. Reprinted with permission from Seemann et al. (2005). Copyright by IOP

From the point of wetting theory, the greatest advantage of these systems has been the ability to study the different dewetting modes, nucleation and spinodal decomposition, although the story has been, in part, a confusing one. The first paper to describe experiments on the dewetting of thin polymer films is due to Reiter (1992). Reiter observed film rupture through the development of holes and interpreted his results in terms of spinodal dewetting. Later work by Jacobs et al. (1998a) showed that, by a statistical sample analysis based on integral geometry (see Appendix B: Minkowski measures), the rupture process was more likely to be based on *heterogeneous nucleation*, see Fig. 3.4. (We will come back to this type of analysis in the context of dewetting dynamics in Chap. 4.)

In the previous chapter we have discussed the case of *homogeneous nucleation* of holes which is driven by thermal energy. In every real system, defects are invariably present. A defect is an object of lower dimension than the system dimension: a surface is a defect to a bulk system. This is why nucleation of droplets from a metastable vapour at a wall is preferred to nucleation in the bulk: nucleation at the surface requires less volume for the critical nucleus provided it does not disfavour

Fig. 3.5 Reconstructed effective interface potential $\Phi(h)$ for PS films on three types of Si wafers. The *cross-hatched area* indicates the experimental error on the determination of the potential minimum by X-ray and contact angle measurements. Reprinted with permission from Seemann et al. (2001b). Copyright by the American Physical Society

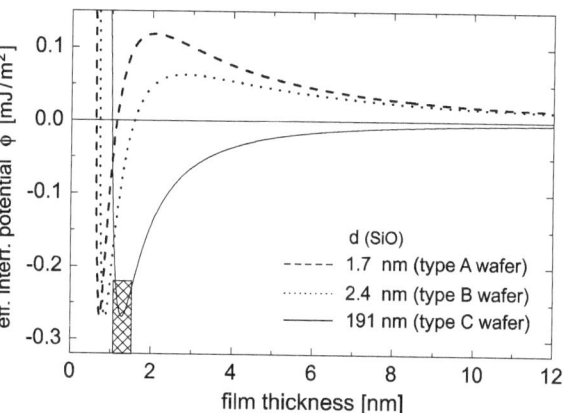

the nucleating phase. Our case of droplet nucleation at $T < T_w$, $\Delta\mu = +0$ is thus properly a heterogeneous nucleation process.

For the quasi two-dimensional wetting films, hole nucleation is first of all a homogeneous (thermal) process, but the free energy barrier can be lowered by lower-dimensional defects, and this appears to be the case for polymer thin films as we will see, although the precise nature of the defects has never been really clarified.

A solution to the problem—dewetting by nucleation or spinodal dewetting—appeared only when it became feasible to produce samples in a very controlled fashion, and to compare experiments very carefully with theoretical concepts based on the effective interface potential. An excellent example for such studies is the work by Seemann et al. (2001b, 2001c, 2005). In this work, polystyrene films were placed on three different wafers whose effective interface potential could be reconstructed by a combination of experimental measurements and calculations (of the Hamaker constant). In the construction of the systems is turned out to be essential to modify the oxide layer on the Si wafer, since this degree of freedom allows to tune the structure of the effective interface potential near the wall. The resulting effective potential is shown in Fig. 3.5. The reconstructed interface potential is given by

$$\Phi(h) = \frac{c_i}{h^8} - \frac{A_{SiO}}{12\pi h^2} + \frac{A_{SiO} - A_{Si}}{12\pi (h + d_i)^2} \tag{3.7}$$

where c_i, $i = A, B, C$ denotes the strength of the short-range interaction on the respective wafer type: $c_A = 1.8(1) \times 10^{-77}$ J/m^6, $c_B = 5.1(1) \times 10^{-77}$ J/m^6, $c_C = 6.3(1) \times 10^{-77}$ J/m^6. The SiO oxide layer thickness d_i for the different wafers is indicated in Fig. 3.5. The two Hamaker constants are given by $A_{SiO} = 2.2(4) \times 10^{-20}$ J and $A_{Si} = -1.3(6) \times 10^{-19}$ J (Seemann et al. 2001b). Exemplary experimental outcomes are shown in Fig. 3.6. It is clearly seen that the barrier of the effective potential is absent for wafer C, while there is a fairly flat barrier for wafer B, and a pronounced barrier for wafer A.

The observed dependencies can nicely be assembled in the stability diagram for PS films as a function of oxide layer thickness in Fig. 3.7. Experimental observations are indicated by the squares, triangles and spheres, corresponding to the previous diagram. From this graph it becomes evident that the process of thermal nucleation is

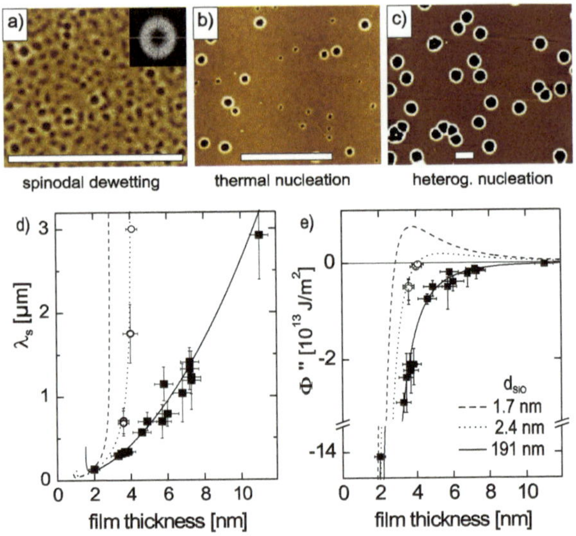

Fig. 3.6 AFM images of dewetting of PS on these wafers. The scale bar in figures (**a**)–(**c**) is 5 μm; the height scale ranges from 0 nm (*black*) to 20 nm (*white*). (**a**) 3.9 nm PS on wafer C; (**b**) 4.1 nm PS on wafer B; (**c**) 6.6 nm PS on type B. (**d**) Spinodal wavelength on wafers B (*open circles*) and C (*filled squares*). (**e**) Second derivative of effective potential. Reprinted with permission from Seemann et al. (2001b). Copyright by the American Physical Society

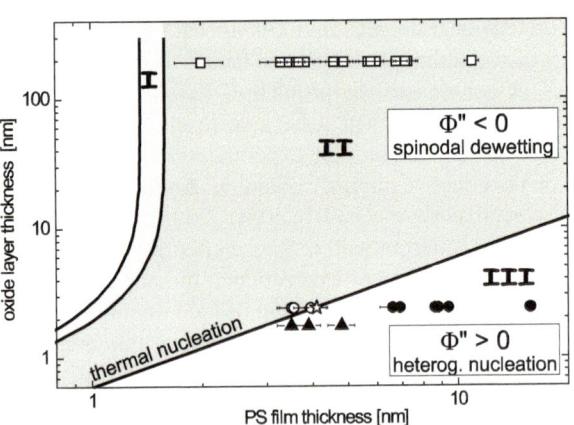

Fig. 3.7 Stability diagram of PS films in the parameters oxide layer thickness and PS film thickness. The three observed behaviours are indicated, as before, by *squares*, *triangles* and *circles*. Reprinted with permission from Seemann et al. (2001c). Copyright by IOP

favoured close to the spinodal $\phi'' = 0$, but farther away from heterogeneous nucleation in which the crossing of the free energy barrier is assisted by some other mechanism, most probably some defect in the film.

A final Task. Compute the critical hole profile $h(r)$ for the metastable polymer film from the first variation of the interface Hamiltonian, Eq. (2.80) with the boundary conditions $h'(0) = 0$ and $h(L) = h_0$, where h_0 is the thickness of the film. Choose L such that the error is sufficiently small. Calculate the excess free energy of the critical hole for the two cases addressed experimentally (wafers A and B).

Part II
Polymer Flow

Chapter 4
Hydrodynamics of Thin Viscous Films

The second Part of the book is devoted to the dynamics of thin polymer films. Although we have studied liquids at surfaces in the first part of the book, the discussion presented there has not made any use of hydrodynamic flow. For the different experimental systems described in Chap. 3, indeed different dynamical models do need to be used. Given the enormous experimental flexibility of polymeric films, we will from now on focus on this case.

In the present chapter we limit ourselves first to the case in which the dynamic behaviour of a thin film is controlled by capillarity, dispersion forces and hydrodynamics. We will also notice that, when pushing into the limit of ultrathin films, we will encounter microscopic properties of polymer films. These will be the topic of the subsequent chapter, which will tackle the viscoelastic properties of thin films, as well as the link between a 'macroscopic' description and microscopics. In this chapter we will derive the basic equations of thin-film hydrodynamics, beginning with the purely viscous case and the so-called weak-slip regime. The derivations are presented in some technical detail but, in order to limit the notational complexity, we will often resort to a two-dimensional setting in which the film thickness is a function of one spatial coordinate along the substrate, and time. The ambition of this first chapter of Part II is to show that, although thin polymer films are controlled by intermolecular forces as they were discussed in Part I, many of their properties can still, and in a highly quantitative fashion, be described by the equations of hydrodynamics as they are used as well for macroscopic fluid flow.

4.1 The Thin-Film Problem

We begin the discussion by specifying the model problem we will analyze in the following. Our starting point is a liquid film of height $z = h(x, y, t)$ deposited on a flat plane, see Fig. 4.1. The liquid is assumed to be incompressible. In a *Eulerian description* the momentum equation is written as (Landau and Lifshitz 1987)

$$\frac{D\mathbf{v}}{Dt} \equiv \partial_t \mathbf{v} + \mathbf{v} \cdot \nabla \mathbf{v} = \nabla \cdot \widehat{\sigma} \tag{4.1}$$

R. Blossey, *Thin Liquid Films*, Theoretical and Mathematical Physics,
DOI 10.1007/978-94-007-4455-4_4, © Springer Science+Business Media Dordrecht 2012

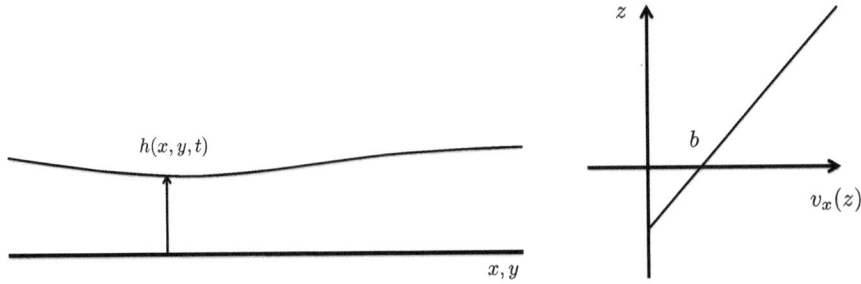

Fig. 4.1 *Left*: Thin film geometry employed throughout the derivations. Note the neglect of over-hangs. *Right*: The velocity profile across the film near the wall to illustrate the Navier slip boundary condition; b is the slip length

where $\mathbf{v}$ is the velocity field of the fluid, D/Dt is the *material derivative*, and

$$\widehat{\sigma} = -p\widehat{1} + \widehat{\tau} \tag{4.2}$$

is the *stress tensor* with its two contributions, the pressure term $-p\widehat{1}$, where $\widehat{1}$ is the $3d$-unit matrix, and $\widehat{\tau}$ as the *extra-stress tensor*, describing shear stresses in the fluid. The form of $\widehat{\tau}$ determines the *fluid model* we wish to use, a point that will occupy us throughout this and the next Chapter. We thus have

$$\frac{D\mathbf{v}}{Dt} = \partial_t \mathbf{v} + \mathbf{v} \cdot \nabla \mathbf{v} = -\nabla p + \nabla \cdot \widehat{\tau} \tag{4.3}$$

as our basic equation. Incompressibility of the fluid leads to the additional condition on the flow field,

$$\nabla \cdot \mathbf{v} = 0. \tag{4.4}$$

Equations (4.3) and (4.4) need to be accompanied by boundary conditions. Since the wall at $z = 0$ is impenetrable the z-component of the velocity field vanishes there,

$$v_z(x, y, z = 0) = 0. \tag{4.5}$$

The flow field components $v_x(x, y, z = 0)$ and $v_y(x, y, z = 0)$ can, in principle, fulfill one of two conditions. The first option is their vanishing at the surface; the second is the *Navier-slip boundary condition*, which is obtained by assuming a velocity profile $v_x(z)$ across the film of the form $v_x(z) = (\partial_z v_x)(z + b)$ near the wall, see Fig. 4.1. Then, the in-plane velocity fields at the wall do not vanish, but obey

$$v_x = b\partial_z v_x \tag{4.6}$$

and likewise for v_y. The length b is called the *slip length*; it is a measure for the friction of the liquid at the surface and will be an important quantity in our further discussions.

For the free interface, $z = h(x, y, t)$, we define a normal vector to the interface as well as the two tangential vectors. From the interfacial surface we remember the definition of g,

$$g \equiv 1 + (\nabla_{\|} h)^2 = 1 + (\partial_x h)^2 + (\partial_y h)^2. \tag{4.7}$$

We then have

$$\mathbf{n} = \frac{1}{\sqrt{g}}(-\nabla_{\parallel}h, 1) \tag{4.8}$$

which is oriented in such a way that it is a *positive* vector for a convex shape (i.e., a droplet-like interface). The tangential vectors read as

$$\mathbf{t} = \frac{1}{\sqrt{g(g-1)}}\left(\nabla_{\parallel}h, (\nabla_{\parallel}h)^2\right) \tag{4.9}$$

and

$$\mathbf{p} = \frac{1}{\sqrt{g-1}}(-\partial_y h, \partial_x h, 1). \tag{4.10}$$

Task: verify that the vectors (4.8), (4.9) *and* (4.10) *are normalized and mutually orthogonal.*

The boundary condition at the free interface is given by

$$\widehat{\sigma} \cdot \mathbf{n} = 2\sigma_{lv}\kappa\mathbf{n} \tag{4.11}$$

where κ is the *mean curvature* of the interface. It is defined by

$$2\kappa \equiv \nabla \cdot \mathbf{n} \tag{4.12}$$

which in components yields the expression

$$2\kappa = -g^{3/2}\left[(\nabla_{\parallel}h)^2 - 2(\partial_x h)(\partial_y h)(\partial_x \partial_y h)\right] \tag{4.13}$$

Task: verify Eq. (4.13).

The normal component of the boundary condition follows from

$$\mathbf{n} \cdot \left(\widehat{\tau} - p\widehat{1}\right) \cdot \mathbf{n} = 2\sigma_{lv}\kappa \tag{4.14}$$

which yields the lengthy relation

$$2\sigma_{lv}\kappa = \left[tr(\widehat{\tau})\right] \cdot (\nabla_{\parallel}h, 1) + 2(\widehat{\tau}_{xz}, \widehat{\tau}_{yz}, \widehat{\tau}_{xy}) \cdot \left(-\nabla_{\parallel}h, (\partial_x h)(\partial_y h)\right) - p \tag{4.15}$$

where $[tr(\widehat{\tau})]$ is the vector whose components are the diagonal elements of the tensor $\widehat{\tau}$. Further, for the tangential components one has the relation

$$\mathbf{t} \cdot \left(\widehat{\tau} - p\widehat{1}\right) \cdot \mathbf{n} = \mathbf{p} \cdot \left(\widehat{\tau} - p\widehat{1}\right) \cdot \mathbf{n} = 0. \tag{4.16}$$

This, explicitly, yields the conditions

$$\left[tr(\widehat{\tau})\right] \cdot \left(-(\nabla_{\parallel}h), (\nabla_{\parallel}h)^2\right) + (\widehat{\tau}_{xz}, \widehat{\tau}_{yz}, \widehat{\tau}_{xy})$$
$$\cdot \left((\partial_x h)\left(1 - (\nabla_{\parallel}h)^2\right), (\partial_y h)\left(1 - (\nabla_{\parallel}h)^2\right), -2(\partial_x h)(\partial_y h)\right) = 0 \tag{4.17}$$

and

$$(\widehat{\tau}_{xz}, \widehat{\tau}_{yz}, \widehat{\tau}_{xy}) \cdot \left((\partial_y h)^2 - (\partial_x h)^2, -\partial_y h, \partial_x h\right) + (\widehat{\tau}_{xx} - \widehat{\tau}_{yy})(\partial_x h)(\partial_y h) = 0. \tag{4.18}$$

Finally, we have a *kinematic criterion* at the free interface which is obtained as follows. We make use of the incompressibility of the film and neglect any exchange with a vapour (this would mean, e.g., a *solvent-free polymer film*). Representing the surface as

$$F(x, y, z, t) \equiv z - h(x, y, z, t) \tag{4.19}$$

the condition that a given material point stays on the interface is then expressed by the condition

$$\frac{DF}{Dt} = 0 \tag{4.20}$$

which is explicitly given by

$$\frac{DF}{Dt} = \partial_t F + (v_x, v_y, v_z) \cdot \nabla F = 0 \tag{4.21}$$

and yields the condition

$$\partial_t h = v_z - (v_x, v_y) \cdot \nabla_{\parallel} h = 0. \tag{4.22}$$

4.2 Lubrication Approximation I

After these preliminaries on the hydrodynamic equations in the thin film geometry, in this section we will derive the *lubrication equation* for thin films by employing tools from *asymptotic analysis*; we follow the approach by Münch et al. (2005).

The essential feature of the analysis employed here is an exploit of the fact that the thin-film geometry allows a separation of length scales—film height vs the transverse dimension of the film, as well as of the lateral and vertical velocity fields in a proper way. It is this insight which permits to reduce the full hydrodynamic equations to a reduced thin-film dynamics.

As we will see the choice of the relationships between length scales and velocity fields *is not unique*. There exist therefore so-called '*distinguished*' *limits* which indeed concern different physical situations, most notably different types of slip strengths. In this first section devoted to the hydrodynamic analysis of thin films, we only discuss the weak-slip case which has been the most-studied case in the literature.

In this first case we consider a purely viscous film. Then, the extra-stress tensor $\widehat{\tau}$ introduced in the previous section is given by

$$\widehat{\tau} = \eta \nabla \mathbf{v}, \tag{4.23}$$

where η is *shear viscosity*. We will see the more general relationships underlying the dynamics of viscoelastic film in Chap. 5.

In the following, we specialize to the case of two dimensions (x, z) in order to reduce the complexity of the derivations; the generalization to the three-dimensional setting is generally fairly straightforward. The flow field is therefore from now on given by

$$\mathbf{v}(x, z, t) = v_x(x, z, t)\mathbf{e_x} + v_z(x, z, t)\mathbf{e_z} \tag{4.24}$$

which fulfills the *Navier-Stokes equation*

$$\varrho(\partial_t \mathbf{v} + \mathbf{v} \cdot \nabla \mathbf{v}) = -\nabla p + \eta \Delta \mathbf{u} \tag{4.25}$$

For thin films, the pressure field is augmented by the *disjoining pressure*, $\Phi'(h)$. As before the fluid is considered as incompressible, hence $\nabla \cdot \mathbf{v} = 0$.

4.2.1 Lubrication Scaling

In the next step we introduce the *lubrication scaling* which will allow us to render the equations non-dimensional. For this we change all variables and functions into capital letters: $x \to X$, $v_x \to V_X$ etc., and then return back to small letters by the definitions:

$$Z \equiv Zz, \qquad X \equiv Xx, \qquad H \equiv Hh, \qquad B \equiv Bb \tag{4.26}$$

while for the velocities we slightly change notation by introducing

$$V_X \equiv V_X u, \qquad V_X \equiv V_X w \tag{4.27}$$

since this allows us to drop the indices. Finally, for time and the pressure field

$$T \equiv \frac{H}{V_Z} t, \qquad P + \Phi' \equiv Pp, \qquad \Phi' \equiv P\varphi' \tag{4.28}$$

The lubrication scaling amounts to assume that (i) the film height H is much smaller than the in-plane length-scale L and, (ii) that the magnitude of the vertical velocity field V_Y is much smaller than the in-plane velocity field magnitude V_X, hence we have identified a small parameter:

$$\frac{H}{L} = \frac{V_Z}{V_X} \equiv \varepsilon \ll 1. \tag{4.29}$$

With this assumption we can write the components of the Navier-Stokes equation in a dimensionless form as

$$\varepsilon \frac{\varrho V_X H}{\eta} (\partial_t u + u \partial_x u + w \partial_z u) = -\varepsilon \frac{PH}{\eta V_X} \partial_x p + \varepsilon^2 \partial_{xx} u + \partial_{zz} u \tag{4.30}$$

$$\varepsilon^2 \frac{\varrho V_X H}{\eta} (\partial_t w + u \partial_x w + w \partial_z w) = -\frac{PH}{\eta V_X} \partial_z p + \varepsilon^3 \partial_{xx} w + \varepsilon \partial_{zz} w \tag{4.31}$$

and the incompressibility conditions as

$$\partial_x u + \partial_z w = 0. \tag{4.32}$$

The boundary conditions at the free interface at $z = h(x, t)$ then are given by

$$\left(\partial_z u + \varepsilon^2 \partial_x w\right)\left(1 - \varepsilon^2 (\partial_x h)^2\right) + 2\varepsilon^2 \partial_x h (\partial_z w - \partial_x u) = 0, \tag{4.33}$$

$$p - \varphi' - 2\varepsilon \frac{\eta V_X}{PH} \frac{(1 - \varepsilon^2 (\partial_x h)^2) \partial_z w - \partial_x h (\partial_z u + \varepsilon^2 \partial_x w)}{1 + \varepsilon^2 (\partial_x h)^2}$$

$$+ \varepsilon^2 \frac{\sigma_{lv}}{PH} \frac{\partial_{xx} h}{(1 + \varepsilon^2 (\partial_x h)^2)^{3/2}} = 0, \tag{4.34}$$

$$\partial_t h - w + u \partial_x h = 0. \tag{4.35}$$

and finally the slip condition

$$u = b\partial_z u. \tag{4.36}$$

This concludes the first step in our analysis: the problem is posed, the equations are non-dimensionalized. In order to proceed we will have to specify how the small parameter ε will be used.

4.2.2 Choice of Scale Ratios

The different scaling regimes emerge from the calculation by balancing the different contributions from pressure, viscosity etc. This can, of course, be done in various ways. In the *weak-slip regime* with which we are concerned right now the pressure gradient is balanced with the dominant viscous term in the horizontal momentum balance, Eq. (4.30), by assuming

$$\frac{PH}{\eta V_X} \sim \varepsilon^{-1}. \tag{4.37}$$

In addition, the liquid-vapour surface tension $\sigma_{lv} \equiv \sigma$ and the pressure balance are assumed to scale according to, see Eq. (4.34)

$$\frac{\sigma}{PH} \sim \varepsilon^{-2}. \tag{4.38}$$

These balances then allow to express the velocity scale in the film as

$$V_X = \frac{\sigma \varepsilon^3}{\eta} \tag{4.39}$$

and the capillary number as

$$Ca = \frac{\eta V_X}{\sigma} = \varepsilon^3. \tag{4.40}$$

Finally, we write the *Reynolds number* as

$$Re = \frac{\varrho V_X H}{\eta} = \varepsilon^3 \frac{\varrho \sigma H}{\eta^2} = \varepsilon^3 Re^*, \tag{4.41}$$

where Re^* is the called the *reduced Reynolds number*, which is an $O(1)$-quantity.

As a result of this operation we obtain the non-dimensional *lubrication problem*

$$\varepsilon^4 Re^* (\partial_t u + u\partial_x u + w\partial_z u) = -\partial_x p + \varepsilon^2 \partial_{xx} u + \partial_{zz} u, \tag{4.42}$$

$$\varepsilon^6 Re^* (\partial_t w + u\partial_x w + w\partial_z w) = -\partial_z p + \varepsilon^4 \partial_{xx} w + \varepsilon^2 \partial_{zz} w, \tag{4.43}$$

$$\partial_x u + \partial_z w = 0, \tag{4.44}$$

and similarly for the boundary conditions, both at the free film surface and the wall which we refrain from writing down here again—and leave it as a task instead.

> Task: write down the non-dimensionalized form of the boundary conditions in the weak-slip regime.

We can now formulate the leading order problem, which is obtained for $\varepsilon = 0$. It is given by

$$\partial_x p = \partial_{zz} u, \tag{4.45}$$

$$\partial_z p = 0, \tag{4.46}$$

$$\partial_x u + \partial_z w = 0, \tag{4.47}$$

with the boundary conditions, first at the free film surface with $z = h(x, t)$,

$$\partial_z u = 0, \tag{4.48}$$

$$p = -\partial_{xx} h + \varphi'(h), \tag{4.49}$$

$$\partial_t h = w - u \partial_x h \tag{4.50}$$

and then at the wall $z = 0$:

$$w = 0, \qquad u = b \partial_z u. \tag{4.51}$$

This leading-order problem can readily be integrated with respect to the vertical coordinate z. Exploiting first that the pressure p does not depend on z (Eq. (4.46)) we find

$$\partial_z u(z) = (\partial_x p)(z - h) \tag{4.52}$$

by enforcing the boundary condition at the free film. Integrating once more over z and employing the boundary condition at the surface we obtain the velocity profile

$$u(z) = (\partial_x p)\left(\frac{z^2}{2} - h(z + b)\right). \tag{4.53}$$

Using finally the kinematic condition at the film surface Eq. (4.22) we obtain the thin-film equation for the weak-slip case,

$$\partial_t h = \nabla\left[M(h)\nabla\left(\frac{\delta \mathcal{H}}{\delta h}\right)\right]. \tag{4.54}$$

where a generalized *mobility*

$$M(h) = \frac{1}{3}h^3 + bh^2 \tag{4.55}$$

has been introduced, while $\mathcal{H}$ is the interfacial Hamiltonian discussed in Chap. 2.

Task: perform the missing steps leading to the final result.

4.3 Mathematical Properties of the Thin-Film Equation

The thin-film equation is highly nonlinear, both due to the mobility $M(h)$ and the interface potential. Further, it is a partial-differential equation of fourth-order. Both these facts render its mathematical properties considerably more complex than, e.g.,

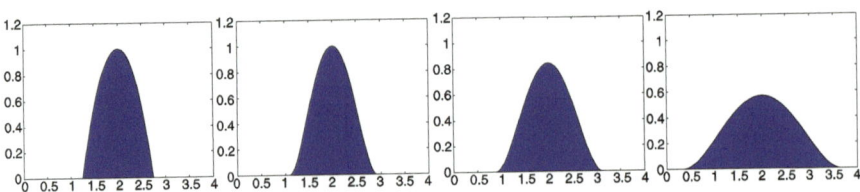

Fig. 4.2 A non-stationary solution to the thin-film equation. Snapshots at $t = 0, 0.001, 0.01, 0.1$. Reprinted with permission from Becker and Grün (2005). Copyright by IOP

that of a diffusion equation. It has therefore attracted wide interest in the mathematical community for which it serves as a laboratory example to further develop the theory of nonlinear partial differential equations.

A first insight into the mathematical complexity can already be gained, following (Becker and Grün 2005), by looking at a simplified problem, a *nonlinear diffusion*

$$\partial_t h + \nabla \cdot \left(M(h) \nabla \Delta h \right) = 0 \tag{4.56}$$

in which only the capillary term is kept, together with initial data

$$h(x, 0) = h_0(x). \tag{4.57}$$

Physically, it corresponds to an interface evolving under surface tension only.

Let us look at the one-dimensional case with $M(h) = h^n$, $n > 0$ on the cylinder $\Omega \equiv (-a, a) \times (0, T)$. The parabolic profile $h(x, t) = M(b^2 - x^2)$ for $b < a$ and $M > 0$ gives a stationary solution to the equation, corresponding to a droplet solution. However, as Fig. 4.2 shows, a non-stationary solution corresponding to a relaxing drop also solves the equation for the same initial data, demonstrating that uniqueness of the solution is not guaranteed.

The origin of the difficulty is clearly an ambiguity in the boundary conditions. Both solutions could be distinguished, e.g., by prescribing the contact angle the liquid makes with the surface. Another version, mostly adopted in the mathematical literature, is to invoke integral relationships on the function and its derivatives. In the case of the thin-film equation with $M(h) = h^n$ and a prescribed constant function as initial data $h(x, T) = h_0$ one has an *energy estimate* (Dal Passo et al. 1998)

$$\frac{1}{2} \int_\Omega |\nabla h(x, T)|^2 + \int_0^T \int_\Omega h^n |\nabla \Delta h|^2 = \frac{1}{2} \int_\Omega |\nabla h_0|^2 \tag{4.58}$$

alongside with a so-called α-*entropy estimate*

$$\frac{1}{\alpha(\alpha + 1)} \int_\Omega h^{\alpha+1}(x, T) + C \int_0^T \int_\Omega \left(\left| \nabla h^{\frac{\alpha+n+1}{4}} \right|^4 + \left| D^2 h^{\frac{\alpha+n+1}{2}} \right|^2 \right)$$
$$\leq \frac{1}{\alpha(\alpha + 1)} \int_\Omega h_0^{\alpha+1} \tag{4.59}$$

provided that $\alpha \in (\max\{-1, \frac{1}{2} - n\}, 2 - n) \neq 0$. Note that the last inequality has no physical interpretation, but only serves to control the solutions in a technical manner. In fact, the one-dimensional parabolic droplet profile from above fulfills the energy estimate, but does not satisfy the entropy estimate. By contrast, the non-stationary profile does fulfill both conditions.

These integral conditions are of great importance in the numerical solution of the thin-film equation in order to guarantee the proper behaviour of the solutions. The calculations presented in the following are based on a scheme developed in Grün and Rumpf (2000). For a brief sketch of the algorithm, see Becker and Grün (2005); and for an even briefer introduction, see Appendix 3.

4.4 Thin Film Rupture in Viscous Thin Films

We now turn to a first application of the thin film equation Eq. (4.54) to the problem of thin film rupture. As we have discussed in detail in Chap. 2 on the statistical mechanics of thin films, there are two possibilities for the film to rupture: the film is either metastable, in which its rupture is brought about by a nucleation process, or it is unstable, in which a process of spinodal dewetting arises; these scenarios are modified in the presence of heterogeneous nucleation mechanisms, as we saw in Chap. 3.

Becker et al. (2003) reported on a detailed combined experimental and computational study of film rupture in the spinodal dewetting regime, following the experimental results on the different dewetting scenarios which we have described in Part I.

In the spinodal dewetting regime, where the effective interface potential has a negative curvature $V''(h_0) < 0$, the dispersion relation obtained from linearizing the thin-film equation in the form $h = h_0 + \delta h$, and assuming $\delta h = h_1 \exp(iqx + \omega t)$ reads as

$$\omega = -q^2 M(h_0)[\sigma q^2 + V''(h_0)] = M(h_0)\Omega(q) \tag{4.60}$$

such that for $q < q_c$ with

$$q_c \equiv \left(-\frac{V''(h_0)}{\sigma}\right)^{1/2} \tag{4.61}$$

a band of unstable modes appear, of which the fastest lies at $d\omega(q)/dq = 0$, which is given by $q_m = q_c/\sqrt{2}$ (see Fig. 4.3).

In this work, PS-films on oxidized Si wafers were studied by atomic force microscopy (AFM). The oxide layer of the substrate had a thickness of 191 nm. The effective interface potential $V(h)$ was expressed as

$$V(h) = \frac{\epsilon}{h^8} - \frac{A_{SiO}}{12\pi h^2} \tag{4.62}$$

where ϵ denotes the strength of the short-range part of the potential with a value of $\epsilon = 6.3(1) \times 10^{-76}$ Jm^6, and A_{SiO} is the Hamaker constant of PS on SiO, $A_{SiO} = 2.2(4) \times 10^{-20}$ J, with the error being given by the bracketed number. The short-range part of potential was constructed in order to fit to the experimentally

Fig. 4.3 Sketch of the
dispersion relation of the
thin-film equation

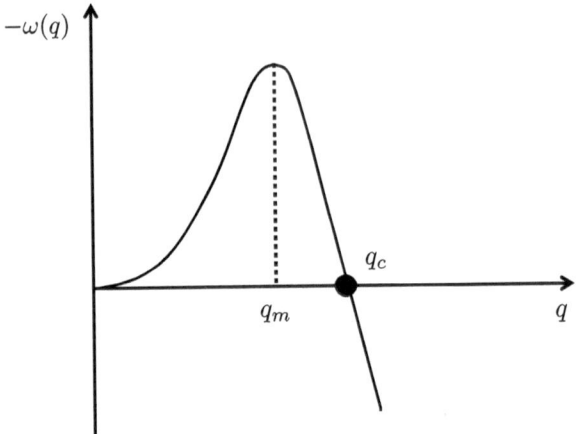

measured location of the global minimum in $V(h)$ obtained from X-ray reflectometry to lie at $h = 1.3(1)$ nm. For the surface tension, a value of $\sigma_{lv} = 30.8$ mNm^{-1} was used. Viscosity is temperature-dependent, so that the values in Figs. 5.3 and 5.4 are, respectively, $\eta = 12{,}000$ Pa s and $\eta = 1{,}200$ Pa s.

The difference between the two Figures lies in the thickness of the initial film. In Fig. 4.4 the film is 3.9(1) nm thick, whereas in Fig. 4.5, it is 4.9(1) nm thick. In both cases, the film rupture is by spinodal dewetting, but in the thicker film, heterogeneous nucleation of holes pre-empts the onset of the linear instability and leads to characteristic satellite hole patterns.

The top graphs in both figures show the results of the AFM scans, while the bottom graphs display the results from numerical simulations of the thin film equation. Several points are to be noted from these graphs, which we will discuss in what follows:

(i) The thin film equation gives a qualitatively correct picture of film rupture. We will show how quantitative the comparison can be made;

(ii) The figures display a difference in the time-scales between experiment and theory. This points to a physical mechanism which is not correctly taken into account. We will see that this discrepancy is due to a neglect of fluctuations in the film;

(iii) Apart from the rupture condition itself, distinct features arise in the holes that form, most notably the shape of the fluid rims. It will turn out that both the substrate properties (slip) as well as viscoelastic properties of the polymers play a role in the formation of these structures.

The analysis of the AFM and simulation data can be performed in a quantitative fashion by making use of the method of *Minkowski measures* (Mecke and Stoyan 2000); see Appendix B. A *Minkowski functional* $M_\nu(A)$ of a pattern A in d dimensions can be defined as an integral over the pattern boundary ∂A via

$$M_\nu(A) = \int_{\partial A} dx\, C_\nu(x) \tag{4.63}$$

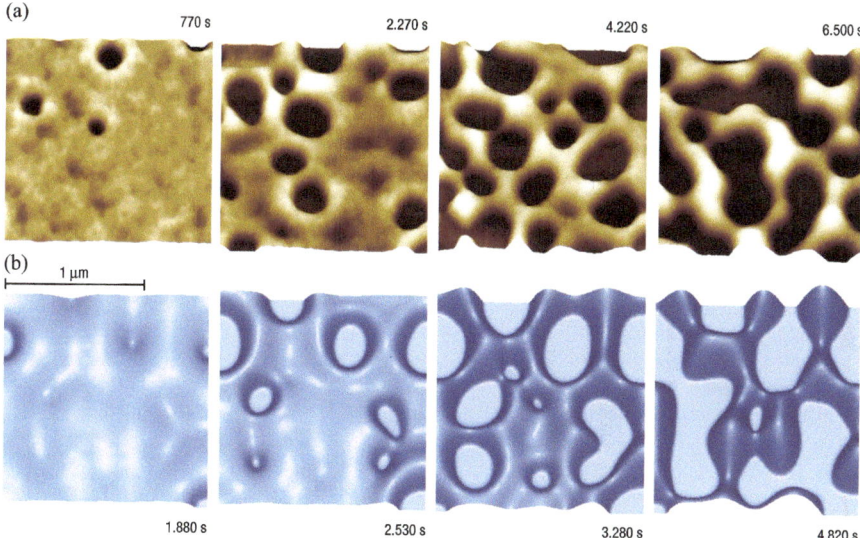

Fig. 4.4 Dewetting morphology of a thin film. (**a**) Experiment: Temporal series of AFM scans recorded in situ at $T = 53$ °C; a 3.9 nm PS film beads off an oxidized Si wafer. (**b**) Simulated dewetting morphology with the identical system parameters as in the experiment. The highest points reach 12 nm above hole ground. The simulation started with a slightly perturbed film. Reprinted with permission from Becker et al. (2003). Copyright by Nature Materials

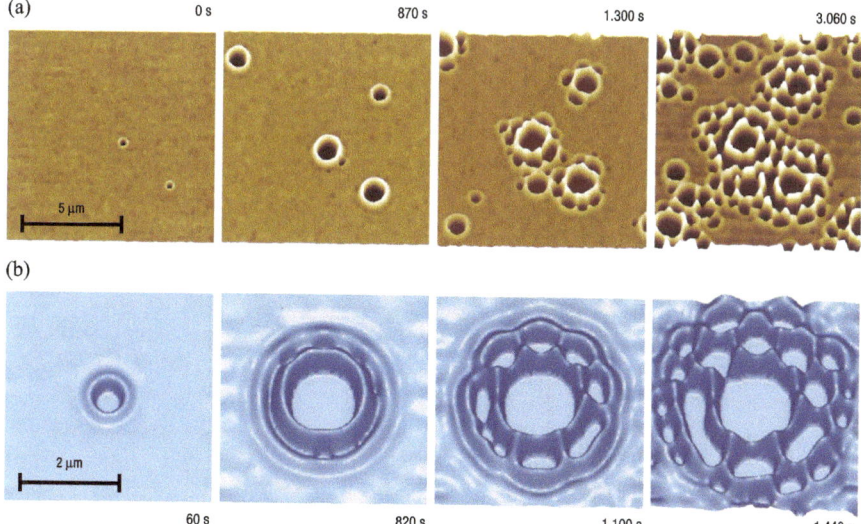

Fig. 4.5 Satellite holes. (**a**) Experimental dewetting scenario of a 4.9 nm PS film on an oxidized Si wafer; temporal series of AFM scans recorded in situ at $T = 70$ °C. (**b**) Simulated scenario for a system with identical properties as in the experiment. Highest points reach 12 nm above hole ground. As initial data a slightly corrugated film with a depression at its centre was chosen. Reprinted with permission from Becker et al. (2003). Copyright: Nature Materials

Fig. 4.6 Analysis of experiment and simulation results by Minkowski measures. Time-averaged and normalized Minkowski measures as a function of threshold value l: (**a**) $s(l)$, (**b**) $u(l)$, (**c**) $\kappa(l)$. (**d**) Ratio $s_0 \kappa_1 / u_2$ as function of time, which is constant for Gaussian fields, although the individual functions vary by two orders of magnitude. Reprinted with permission from Becker et al. (2003). Copyright: Nature Materials

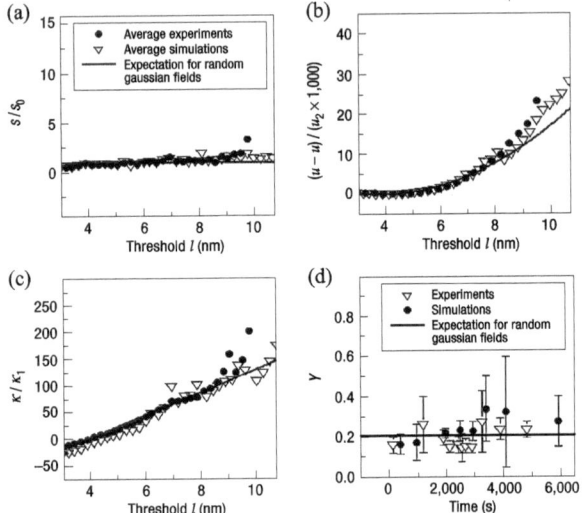

where, e.g., in two dimensions $C_0(x) = x \cdot n/2$ with n as the normal vector at each point of the boundary, $C_1(x) = 1$, and $C_2(x)$ is curvature. The corresponding measures of a threshold image, as used here, where $h > l$ are related to *area* $M_0 = F(l)$, *boundary length* $M_1 = U(l)$ and the *Euler characteristic* $M_2 = \chi(l)$. For convenience in the analysis, composed quantities have been defined which are related to the *fundamental measures* via the following relations:

$$s(l) \equiv -\frac{F(l)}{U(l)} \tag{4.64}$$

is an effective slope,

$$u(l) \equiv \ln U(l) \tag{4.65}$$

is a logarithmic boundary length and

$$\kappa(l) \equiv \frac{\chi(l)}{U(l)}. \tag{4.66}$$

Based on these quantities, the surfaces are characterized by contour lines given by isosurfaces $h(x,t) = l$. Figure 4.6 displays the above-defined Minkowski measures as functions of threshold l. The quantitative accord between experiments and theory is striking. Both experiments and simulations follow a *Gaussian random-field model* for contours above the average film thickness $l_0 \approx 3.9$ nm, namely $s = s_0$, $u(l) = u_0 - u_2(l - l_0)^2$, and $\kappa(l) = \kappa_1(l - l_0)$. In the figure, only time-averaged data are shown, but similar accord is found for each snapshot at any given time; only the parameters s_0, u_2 and κ_1 have an algebraic decay on time t with exponents $2v_s \approx 2v_\kappa \approx v_u \approx 1.8 \pm 0.2$ over at least two decades. A convincing consistency check is shown in figure d, where the ratio

$$Y \equiv \frac{s_0 \kappa_1}{u_2} = \frac{2}{\pi^2} \approx 0.203 \tag{4.67}$$

is plotted, which stays constant for a Gaussian random field (Becker et al. 2003).

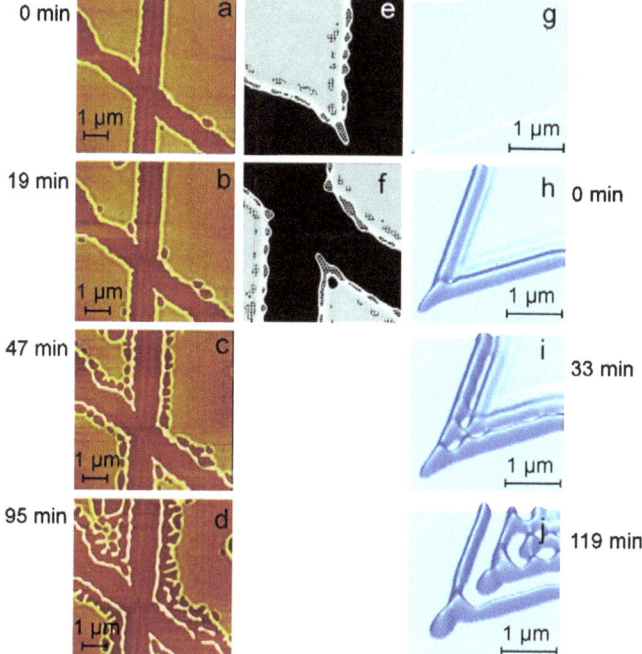

Fig. 4.7 *Top*: dewetting in a wedge geometry, comparison of AFM data (*left*, **a–d**) and simulations (*right*, **g–j**). The *middle images* (**e, f**) are magnifications from the corresponding AFM images to the *left*. Reprinted with permission from Neto et al. (2003). Copyright by IOP

The quality of the agreement between simulation and experiment is further qualitatively illustrated by Fig. 4.7 which shows the dewetting of the polymer film in a wedge geometry.

4.5 Thermal Fluctuations in the Film

The excellent accord between theory and experiment discussed before has one remaining deficiency which is the discrepancy of the value of the viscosity, or equivalently, the correct time-scale for rupture. It is not to be expected that, at this level, viscoelastic properties will intervene (see Chap. 5).

Within the present theory, thermal fluctuations of the film height—thermally excited capillary waves—are neglected. The role that such fluctuations can play in the film dynamics has been analyzed theoretically (Mecke and Rauscher 2005, Grün et al. 2006) and in comparison to experiment (Fetzer et al. 2007a, 2007b).

In a first step, the thin film equation has to be extended to include thermal noise. Prescriptions on how this can be done can be taken from fluctuating bulk hydrodynamics (Landau and Lifshitz 1987), or applications to jets (Moseler and

Landmann 2000) and (Eggers 2002). The thin-film equation then acquires the form

$$\eta \partial_t h = \nabla \cdot \left(\frac{h^3}{3} \nabla \left[-\sigma_{lv} \Delta h + \Phi'(h) \right] + D(h) \zeta(t) \right) \tag{4.68}$$

where $D(h) \equiv \sqrt{(2/3)k_B T h^3 \eta}$ and the noise is given by a stochastic process with vanishing average

$$\langle \zeta(x, t) \rangle = 0 \tag{4.69}$$

and correlations

$$\langle \zeta_i(x, t) \zeta_j(x', t') \rangle = \delta_{ij} \delta(x - x') \delta(t - t') \tag{4.70}$$

where $\langle \cdot \rangle$ is the ensemble average over the noise realizations.

The above described experiments and simulations were reanalyzed based on these equations in Fetzer et al. (2007a, 2007b). First, we can get an idea of the magnitude of the expected effect. For the two experiments, the characteristic length scales are as follows. For an initial film height of about $h_0 = 4$ nm, the spinodal wavelength is given by

$$\lambda = \frac{2\pi}{q_0} = \sqrt{-8\pi^2 \sigma_{lv}/\Phi''(h_0)} = 4h_0^2 \sqrt{\pi^3 \sigma_{lv}/A} \tag{4.71}$$

for an effective interface potential $\Phi(h) = -A/(12\pi h^2)$. For $A \approx 2 \cdot 10^{-20}$ Nm and $\sigma_{lv} \approx 3 \cdot 10^{-2}$ N/m one obtains $\lambda \approx 400$ nm. These parameters lead to approximate noise amplitudes of the order of

$$\frac{3k_B T}{8\pi^2 h_0^2 \sigma_{lv}} \approx 4 \cdot 10^{-4} \tag{4.72}$$

and $2 \cdot 10^{-2}$, respectively, for both experiments.

Quantitative information on the influence of thermal noise can be obtained by considering the variance of the film height $\sigma^2(t) = \overline{h^2} - \overline{h}^2$ and, in addition, the variance of the local slope of the film height $h(x, t)$, which is given by

$$k^2(t) = \frac{\overline{(\nabla h)^2}}{2\pi \sigma^2(t)}. \tag{4.73}$$

This quantity contains information about the preferred wavevector in the film surface during the early stages of the dewetting process. The overbar indicates an integration over all positions in the images.

Since both the quantities $\sigma^2(t)$ and $k^2(t)$ will be determined for the linear behaviour of the thin film equation (which means either early or later stages of the dewetting process), the observation window is, however, limited both in overall size and due to the appearance of growing holes, the analysis has to be based on the real-space images.

Figures 4.8(a) and (b) display the analysis of the AFM experiments and the deterministic simulations. For $\sigma^2(t)$, the agreement between experiments and deterministic ($T = 0$) simulation is satisfactory, but this is not the case for $k^2(t)$. This quantity

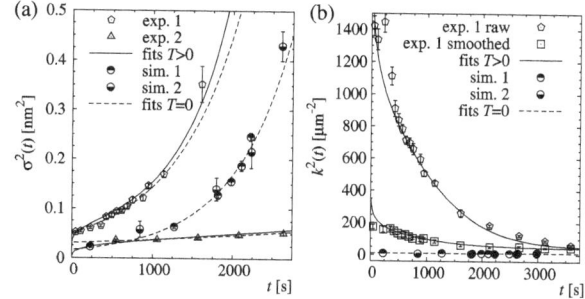

Fig. 4.8 Noise analysis of AFM experiments and of deterministic simulations. (**a**) $\sigma^2(t)$; (**b**) $k^2(t)$. For the discussion, see main text. Reprinted with permission from Fetzer et al. (2007a, 2007b). Copyright by the American Physical Society

stays constant for the deterministic simulations, but clearly varies with time for the experiment.

In order to compare the experiment with the theory comprising thermal fluctuations, an analysis of for $\delta h(x,t) \equiv |h(x,t) - h_0| \ll h_0$ has been performed for which the thin-film equation can be linearized and the capillary wave spectrum computed exactly, with the result

$$\tilde{C}(q,t) = \langle \delta\tilde{h}(q,t)\delta\tilde{h}^*(q,t)\rangle$$

$$= \tilde{C}_0(q)\exp\left(2\omega(q)t\right) + \frac{k_B T h_0^3}{3\eta}\frac{q^2}{\omega(q)}\left[\exp\left(2\omega(q)t\right) - 1\right] \quad (4.74)$$

with the dispersion relation

$$\omega(q) = \frac{1}{t_0}\left[1 - \left(\frac{q^2}{q_0^2} - 1\right)^2\right] \quad (4.75)$$

where $q_0^2 = -\Phi''(h_0)/(2\sigma_{lv})$ is the maximum of the dispersion relation and $t_0 = 3\eta/(\sigma_{lv}h_0^3 q_0^4)$ is a characteristic time scale, $t_0 = 300$ s for experiment 1 and $t_0 = 1700$ s for experiment 2. The initial spectrum $\tilde{C}_0(q)$ at $t = 0$ is assumed constant, $\tilde{C}_0(q) = 2\pi^2\sigma_0^2/q_0^2$ for $q < \sqrt{2}q_0$ and zero otherwise. In addition, a microscopic cutoff $q_m = 2\pi/r_m \gg q_0$ is imposed, whereby r_m is set by the pixel size resolution. The noise that builds up in short wavelength roughness is irrelevant for the film evolution on the timescale $t_0 \gg t_m = (q_0/q_m)^4 t_0$.

Figure 4.9 compares the experimental data to fits to the linearized stochastic thin film equation. It is clearly seen that smoothing of the data has a significant effect; smoothing removes height fluctuations of the film on small scales, which in experiment are regenerated all the time. Also shown is a comparison with an asymptotic formula obtained for $t > t_m \approx 10^{-2}t_0$ up to t_0 where one has (Mecke and Rauscher 2005)

$$k^2(t) \approx k_0^2(t) + \frac{\chi}{\sigma^2(t)}, \qquad \chi = \frac{k_B T}{8\pi^2\sigma_{lv}}q_m^2, \quad (4.76)$$

independent of the initial conditions (see insert). Again the agreement is very good, full consistency between experiment and theory is obtained. In particular, the microscopic length scale r_m as extracted from the fits, lies in the range of about 100 nm, as expected.

Fig. 4.9 Slope variance k^2 from experiments vs roughness σ^2. The *solid lines* are fits based on the linearized stochastic thin-film equation, the *dashed-dotted lines* asymptotic fits to formula (4.73). Reprinted with permission from Fetzer et al. (2007a, 2007b). Copyright by the American Physical Society

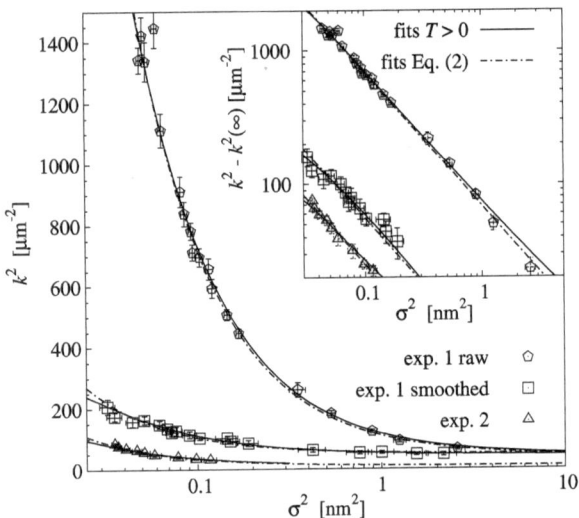

4.6 Lubrication Approximation II

We now turn to further distinguished limits of the thin-film problem, the so-called *strong-slip* limit and the *intermediate slip* case (Münch et al. 2005). In the *strong-slip* case, Eq. (4.37) is replaced by the leading-order assumption

$$\frac{PH}{\eta V_x} \sim \varepsilon, \tag{4.77}$$

which balances the vertical momentum balance against pressure in Eq. (4.31). The surface tension vs. pressure balance remains unchanged, as does the balance of pressure against surface tension, but the scales for the velocity and capillary number are modified according to

$$V_x = \frac{\sigma \varepsilon}{\eta} \tag{4.78}$$

and

$$Ca = \frac{\eta V_x}{\sigma} = \varepsilon. \tag{4.79}$$

The Reynolds number is

$$Re = \frac{\varrho V_x H}{\eta} = \varepsilon \frac{\varrho \sigma H}{\eta^2} = \varepsilon Re^*. \tag{4.80}$$

Consequently, these quantities are now of order ε, i.e., much more relevant for the dynamics than in the weak-slip case—a fact which will show up in the final equations.

As for the weak-slip regime, we can exploit these scalings for the Navier-Stokes and conservation equations, as well as for the boundary terms, a calculation we do not present here. It turns out that the lowest order terms in the lubrication parameter

ε appear to $O(\varepsilon^2)$ so that one has to assume the existence of asymptotic expansions for the functions $u(x, z, t; \varepsilon)$, $w(x, z, t; \varepsilon)$, $p(x, z, t; \varepsilon)$ and $h(x, t; \varepsilon)$ in ε^2. The resulting leading order problem then has the form

$$\partial_{zz} u_0 = 0, \tag{4.81}$$

$$\partial_z p_0 = \partial_{zz} w_0, \tag{4.82}$$

$$\partial_x u_0 + \partial_z w_0 = 0, \tag{4.83}$$

with the boundary conditions at the interface $z = h_0(x, t)$:

$$\partial_z u_0 = 0$$
$$p_0 - \varphi' - 2(\partial_z w_0 - \partial_x h_0 \partial_z u_0) + \partial_{xx} h_0 = 0 \tag{4.84}$$
$$\partial_t h_0 - w_0 + u \partial_x h_0 = 0.$$

At the surface $z = 0$ we have

$$w_0 = 0, \qquad \partial_z u_0 = \frac{u_0}{b}. \tag{4.85}$$

From Eq. (4.81-1), the leading-order horizontal velocity is independent of z, $u_0 \equiv u_0(x, t)$. From Eqs. (4.81, 4.84-3) we find $w_0 = -z \partial_x u_0$ while from Eq. (4.84-2), one finds

$$p_0 - \varphi' = -\partial_{xx} h_0 - 2 \partial_x u_0. \tag{4.86}$$

At this point we have to decide on the dependence of the slip length on the lubrication parameter. In the weak-slip regime, the slip length b was assumed as an $O(1)$-quantity. Then, the above equations imply that $u_0 = w_0 = \partial_t h_0 = 0$, which is obviously not sufficient. We therefore have to allow

$$b = \frac{\beta}{\varepsilon^\alpha} \tag{4.87}$$

with $\alpha > 0$. This corresponds to a choice of $b \gg O(1)$, hence indeed to a large-slip regime.

To leading order, we now have $\partial_z u_0 = 0$, and the zero-order problem is satisfied. We thus have to go to the next-order correction in which no explicit dependence on ε arises:

$$Re^*(\partial_t u_0 + u_0 \partial_x u_0) = -\partial_x p_0 + \partial_{xx} u_0 + \partial_{zz} u_1, \tag{4.88}$$

$$Re^* z(\partial_{xt} u_0 + u_0 \partial_{xx} u_0 + (\partial_x u_0)^2) = -\partial_z p_1 + z \partial_{xxx} u_0 + \partial_{zz} w_1, \tag{4.89}$$

$$\partial_x u_1 + \partial_z w_1 = 0, \tag{4.90}$$

with the boundary conditions at the free film surface $z = h(x, t)$, which turn out to be somewhat lengthy expressions:

$$\partial_z u_1 - \partial_{xx} h_0 - 4 \partial_x h_0 \partial_x u_0 = 0, \tag{4.91}$$

$$p_1 - \varphi''(h_0) h_1 - 2(\partial_x u_0 (\partial_x h_0)^2 + \partial_z w_1 - \partial_z u_1 \partial_x h_0 + h_0 \partial_x h_0 \partial_{xx} u_0)$$
$$- 2 \partial_x u_0 (\partial_x h_0)^2 + \partial_{xx} h_1 - \frac{3}{2} (\partial_x h_0)^2 \partial_{xx} h_0 = 0, \tag{4.92}$$

and

$$\partial_t h_1 - w_1 + u_0 \partial_x h_1 - u_1 \partial_x h_0 = 0. \tag{4.93}$$

The boundary condition at the wall surface $z = 0$ is then given by

$$w_1 = 0, \qquad \partial_z u_1 = \frac{\varepsilon^{\alpha-2}}{\beta} u_0. \tag{4.94}$$

We see that the distinguished limit corresponds to $\alpha = 2$ in this case. Integration of Eq. (4.89) over z across the film from 0 to h_0, and using Eqs. (4.91) and (4.94-1), one finds the strong-slip equations for the thin film

$$h_0 Re^* (\partial_t u_0 + u_0 \partial_x u_0) = 4 \partial_x (h_0 \partial_x u_0) + h_0 \partial_x (\partial_{xx} h_0 - V'(h_0)) - \frac{u_0}{\beta} \tag{4.95}$$

$$\partial_t h_0 = -\partial_x (h_0 u_0). \tag{4.96}$$

These two equations are the *strong-slip model for Newtonian thin films*. Apart from Münch et al. (2005) it had been derived independently by Kargupta et al. (2004).

The strong-slip case is as rich as the weak-slip case. While the latter allows for the limit $b \to 0$, hence the no-slip case, the former allows exponent values $\alpha > 2$, leading naturally to the lubrication model for free-standing films. In this case, the slip term in Eq. (4.95) is absent.

We now turn to the intermediate regime. In this case, we allow for the scalings

$$V_x = \frac{\sigma \varepsilon^{2-\delta}}{\eta}, \tag{4.97}$$

$$Ca = \varepsilon^{2-\delta} \tag{4.98}$$

and

$$Re = \varepsilon^{2-\delta} Re^* \tag{4.99}$$

for $-1 < \delta < 1$. Again we do not enter the details of the calculation in this case and refer to the paper by Münch et al. (2005). The final result can be represented as a single lubrication equation

$$\partial_t h + \beta \partial_x \left[h^2 \partial_x (\partial_{xx} h - \phi'(h)) \right] = 0 \tag{4.100}$$

which corresponds to the *slip-dominated* lubrication model, in which the no-slip term is omitted from the weak-slip equation.

> *Task. Perform a linear stability analysis for a flat film obeying the intermediate and strong-slip thin film dynamics. Write down and discuss the dispersion relations. See Rauscher et al. (2008) for the strong-slip case.*

4.7 Dewetting Holes

In this section, we will disentangle three aspects of dewetting holes:

- the time-evolution of the hole radius;

- the theory of the dewetting rim, first without and then with slip;
- the instability of the rim.

We begin with the growth law for a dewetting hole.

4.7.1 Hole Growth

The growth law of a dewetting hole is not easy to extract from the full dynamical equations, other than by numerical means (which of course has been done). It is, however, possible to derive an approximate analytic formula which captures the observed behaviour very well and has the advantage to allow to gain some more insight (Jacobs et al. 1998b).

Starting point for the derivation of this formula is the viscous dissipation associated with the rim of an opening hole. Making use of the *vorticity*

$$\omega = \nabla \times \mathbf{v} \tag{4.101}$$

we can write with the *enstrophy*

$$Z = \frac{1}{2}\omega \cdot \omega \tag{4.102}$$

the dissipation of the rim as

$$P_v \equiv \eta \int_{rim} dx dz \omega^2 \tag{4.103}$$

where we have considered a quasi two-dimensional situation with the (x, z)-cross section of the rim; this is permissible when the rim width w is small compared to the hole radius R. We express this relation as

$$P_v \equiv v_v^2 K_v(\theta) \tag{4.104}$$

where v_v is the flow velocity at the three-phase contact line, and $K_v(\theta)$ a geometric factor whose essential dependence is on contact angle.

In the situation of flow, the driving force is the nonequilibrium spreading coefficient S, since it governs the gain in surface free energy per unit area that results from film retraction. Consequently, our dynamic equation is

$$S\dot{R} = Sv_v = v_v^2 K_v(\theta) \tag{4.105}$$

and we immediately obtain the viscous growth law

$$R(t) = \frac{S}{K_v}t \tag{4.106}$$

valid in the no-slip case.

In the case of slip, we have to replace the velocity by the expression

$$v_{slip}(x) = v_s f\left(\frac{x}{w}\right) \tag{4.107}$$

Fig. 4.10 *Dashed line*: theoretical prediction for no-slip conditions. *Dotted line*: prediction for full slip, i.e., when dissipation takes place only by slippage of the film across the substrate. *Solid line*: Eq. (4.114) fitted to the data, which allows to extract the effective viscosity and the slip friction of the liquid film on the substrate. Reprinted with permission from Seemann et al. (2005). Copyright by IOP

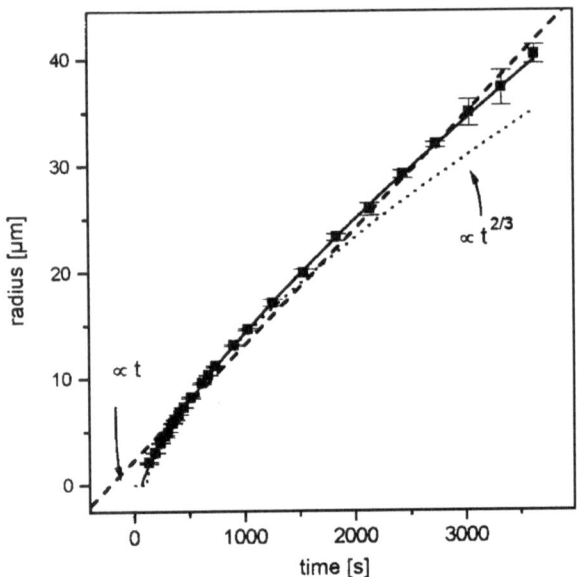

where $f(0) = 1$ and $f(y) \approx 0$ for $y > 1$. If f is not dependent on time, which is assumed, we can estimate the dissipation integral as

$$P_{slip} = \overline{\zeta} v_s^2 w \qquad (4.108)$$

where $\overline{\zeta} = \int_0^w dx f(x/w)$, where ζ is the friction coefficient between the film and the substrate. Then, in analogy to the no-slip case, we find

$$S = \overline{\zeta} v_s w \equiv v_s K_s \sqrt{R} \qquad (4.109)$$

where we have used the fact that due to conservation of mass, $w \sim \sqrt{R}$. For the receding front with

$$\dot{R} = v_s \qquad (4.110)$$

we then find the growth law

$$R(t) = \left(\frac{3S}{2K_s} t \right)^{2/3}. \qquad (4.111)$$

Indeed, the two extreme cases of no-slip and pure slip can be combined into the equation

$$\dot{R} = v_s + v_v \qquad (4.112)$$

which superimposes the two dissipation mechanisms, viscous and due to the substrate. We then have in terms of the radius R

$$\dot{R} = S \left(\frac{1}{K_v} + \frac{1}{K_s} \sqrt{R} \right) \qquad (4.113)$$

which can be integrated by separation of variables and leads to an analytic expression for $t(R)$ in the form

$$t = \frac{K_v}{S}\left(R - 2\alpha\sqrt{R} + 2\alpha^2 \ln\left(1 + \frac{\sqrt{R}}{\alpha}\right)\right) + \tau_0 \tag{4.114}$$

where $\alpha \equiv K_v/K_s$.

Figure 4.10 plots the theoretical curves in comparison with experimental data obtained for PS films with molecular weights M_w between 18 and 600 kg/mol, with a ratio of $M_w/M_n = 1.02$ and a $T_g = 100\,°C$ (see Appendix A). The agreement is satisfactory.

4.7.2 The Dewetting Rim

As shown in Fig. 4.5, the rupturing film displaces the liquid, creating a liquid rim surrounding the developing hole: this is nothing but a simple consequence of mass conservation. However, as it turns out, the detailed experiments that can be performed with AFM on the rupturing films have shown that the radial profile of the rim displays features that are a direct consequence of the underlying thin-film dynamics. The following two figures exemplify these results. They show two different behaviours, one in which the rim profile decays monotonously towards the film, while in the other graph, the profile towards the rim decays with a small depression. This feature of a rupturing thin film has created a still ongoing discussion in the literature with regards to its origin, in particular, which role play slip and/or viscoelasticity in the film.

Figure 4.11 bottom defines a number of length scales associated with the dewetting rim which have been employed in a simplified first model for the rim. Indeed, by linearizing the thin-film equation around the flat film value $h = h_0$ and going to a co-moving frame with the velocity U of the contact line, one finds (Seemann et al. 2001d)

$$\partial_t h + \frac{\sigma_{lv} h_0^3}{3\eta}\partial_{xxxx}h - v\partial_x h = 0, \tag{4.115}$$

neglecting the contribution from the disjoining pressure. The traveling wave solution in $\xi \equiv x - Ut$ then is

$$h(\xi) = \exp\left(-\frac{2\pi}{\sqrt{3}\ell}\right) \cdot \cos\left(\frac{2\pi}{\ell}\xi\right) \tag{4.116}$$

where the lengthscale ℓ is defined as

$$\ell \equiv \frac{4\pi}{\sqrt{3}h_0} \cdot \left(\frac{\sigma_{lv}}{3\eta U}\right)^{1/3}, \tag{4.117}$$

hence a damped harmonic oscillation. The ratio of $W/V \approx 0.016$ differs from the experimental value by a factor 2, which can be improved by including a boundary condition at the rear angle α.

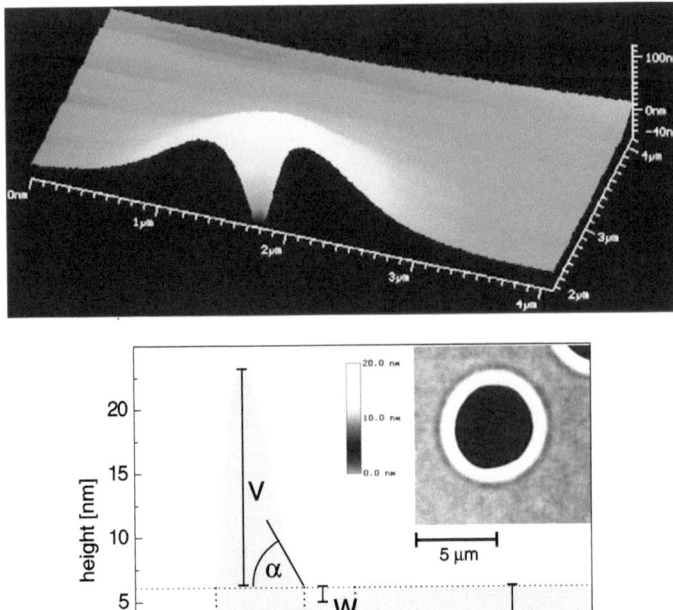

Fig. 4.11 *Top*: AFM cross section of a hole in a PS(65k) film on OTS-Si (for an explanation of these experimental terms, see Appendix A). The material removed from the hole has accumulated in a characteristic rim at the circumference of the hole. Notice the large difference in lateral and vertical scales. *Bottom*: AFM scan of a hole in a 6.6(2) nm thick PS(2.24k) film on an SiO-wafer with an oxide layer of 191 nm. The hole profile displays a well-visible depression W next to the rim. Reprinted with permission from Seemann et al. (2005). Copyright by IOP

It is a simple fact that the linear fourth-order equation easily leads to damped oscillatory profiles. These, are, however, only poor approximations to the real profile. In the next section, we therefore discuss an asymptotic solution for the full profile of the dewetting rim of a weakly slipping thin film during rupture.

4.7.3 Dewetting Rim: Asymptotic Analysis in the No-Slip Case

The full profile of the depression behind the dewetting rim can be calculated from an asymptotic analysis of the nonlinear thin-film equation. We follow here the work by Snoeijer and Eggers (2010).

Figure 4.12 shows a qualitative sketch of the profile in which the relevant parameters are indicated. As in the derivation of the thin film equation we can take

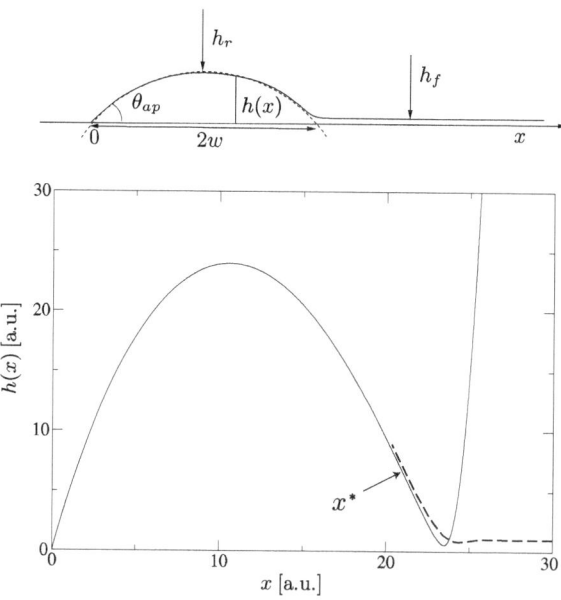

Fig. 4.12 *Top*: Sketch of the dewetting rim profile $h(x)$. A liquid film of thickness h_f is invaded by a moving rim of height h_r and width w. The apparent contact angle θ_{ap} is defined as the intersection of the solid surface and a parabolic fit of the rim (*dashed line*). *Bottom*: Schematic representation of the matching procedure. The contact line and rim are described by the *solid line*, from Eq. (4.124). The *dashed line* represents the profile in the film region, from Eq. (4.125). The profiles are matched at the advancing side of rim around the point x^*. Reprinted with permission from Snoeijer and Eggers (2010). Copyright by the American Physical Society

the film profile as two-dimensional, since for the expanding hole the dewetting rim becomes increasingly straight. Starting point is then the lubrication equation in a frame co-moving with the contact line

$$\partial_t h - U \partial_x h + \frac{\sigma}{\eta} \partial_x \left(\mathcal{M}(h) \partial_{xxx} h \right) = 0. \tag{4.118}$$

with the boundary conditions

$$h(0, t) = 0, \qquad \partial_x h(0, t) = \theta \tag{4.119}$$

while at the film end the profile approaches the flat film value

$$h(x \to \infty) = h_0. \tag{4.120}$$

Since the film mass is conserved, the material accumulates in the rim upon hole expansion. If the film area is given by $A(t) \sim w^2$ where w is the halfwidth of the rim, it will grow according to $\dot{A} = w\dot{w}$. If the advancing of the film is given by Uh_0, one has

$$\frac{\dot{w}}{U} \sim \frac{h_0}{w} \tag{4.121}$$

which becomes asymptotically small for long times. Changes in the geometry are therefore slow with respect to U, allowing to drop $\partial_t h$ when compared with $U \partial_x h$. One thus ends up with the quasi-stationary problem

$$-(Ca)h + \mathcal{M}(h)\partial_{xxx}h = Q \tag{4.122}$$

after integrating once with flux Q as integration constant. $Ca = U\eta/\sigma$ is the capillary number, as in the Introduction. Hence, resolving for $\partial_{xxx} \equiv h'''$,

$$h'''(x) = \frac{3(Ca)h + 3Q}{\mathcal{M}(h)} = \frac{3(Ca)h + 3Q}{h^2(h + 3\lambda)}. \tag{4.123}$$

The boundary conditions have now to be adapted to this equation; clearly, it does not fulfill the original conditions anymore due to the neglect of terms of order h_0/w due to the stationarity assumption. However, the interest here is to match the front problem to a rear problem at a value $x = x^*$. At the contact line $h = 0$ one has for $h = 0$ a flux value of $Q = 0$ so that

$$h''' = \frac{3(Ca)}{h(h + 3\lambda)} \tag{4.124}$$

for $0 \leq x < x^*$, while for the latter we have for $x > x^*$

$$h''' = \frac{3(Ca)}{h^3}(h - h_0) \tag{4.125}$$

where for the last equation we have for $h = h_0$, $h''' = 0$ the flux value $Q_f = -(Ca)h_0$. Moreover, in the last equation, $h \gg \lambda$ has been supposed.

    We are now left with a matching problem. We do not trace the whole derivation of the solution here which can be found in Snoeijer and Eggers (2010), but only give the solution strategy here. The first step consists in the formulation of a suitable non-dimensional problem. Since the only appearing length-scale in Eq. (4.124) is the slip length λ, a rescaling of $h(x)$ can be suggested according to

$$h(x) = 3\lambda H\left(\frac{x\theta}{3\lambda}\right) \tag{4.126}$$

and the definition of $\xi \equiv x\theta/3\lambda$. One then rewrites Eq. (4.124) in the form

$$H'''(\xi) = \frac{3Ca/\theta^3}{H^2 + H}. \tag{4.127}$$

For $h \gg \lambda$, the equation reduces to

$$y''' = \frac{1}{y^2} \tag{4.128}$$

where $H(\xi) = [(3Ca)^{1/3}/\theta]y(\xi)$. This equation has an exact solution involving Airy functions, whose properties can be exploited for the matching analysis, see Snoeijer and Eggers (2010).

    From this analysis follows that the rim has a parabolic shape of the form

$$y(\xi) \propto y_{max}\left[1 - \left(\frac{\xi - \xi_{max}}{\xi_{max}}\right)^2\right] \tag{4.129}$$

and that the profile develops a minimum at a value $\xi_{min} = 2\xi_{max}$. Near the minimum ξ_{min} the solution develops a logarithmic dependence so that one finds

$$y'^3 = -3\ln\left(n\alpha^2(\xi_{min} - \xi)\right) \tag{4.130}$$

where n is a numerical factor and α^2 a ratio of matching constants related to the Airy solutions of Eq. (4.124).

We now turn to the film profile, where the task now is to extract a similar behaviour of the corresponding solution to Eq. (4.125). This is possible with the choice

$$h(x) = h_0 G(\zeta) \tag{4.131}$$

where $\zeta = (x - x_0)(Ca)^{1/3}/h_0$, transforming Eq. (4.125) into

$$G'''(\zeta) = \frac{3}{G^2}\left(1 - \frac{1}{G}\right), \tag{4.132}$$

from which the behaviour

$$G'^3 \propto -9\ln\left(a[\zeta_0 - \zeta]\right) \tag{4.133}$$

is obtained. The matching procedure now involves to relate the constant a in Eq. (4.125) to $n\alpha^2$. The final result is

$$Ca = \frac{\theta^3}{9}(\ln Y)^{-1} \tag{4.134}$$

where $Y \sim (Ca)^{1/3}(w/\lambda h_0)^2$. This concludes the analysis, since both asymptotic solutions are found and matched.

4.7.4 Dewetting Rims in Slipping Films

In this subsection we attempt to demonstrate what new dynamic properties exist for the dewetting rim in the strong-slip regime. One can now, based on the strong-slip and intermediate-slip equation as well perform the asymptotic analysis for the rim, as we did for the weakly slipping film above. Such studies were described in Flitton and King (2004) and Münch et al. (2005); we refer to these papers for the corresponding results. For our purpose here, the interesting fact is that, indeed, for sufficiently large slip length, the oscillatory profile can be turned into a monotonously decaying one.

This can be seen from the following simple argument applied to the strong-slip equation. We linearize the equations around the normalized flat film solution $h(x, t) = 1$ in the form

$$h(x,t) = 1 + \delta\varphi(\xi), \qquad u(x,t) = \delta v(\xi), \qquad \xi = x - s(t) \tag{4.135}$$

where we went into a comoving frame with the position of the contact line given by $s(t)$. The incompressibility condition then transforms into

$$\partial_{xi}(v - \dot{s}\varphi) = 0 \tag{4.136}$$

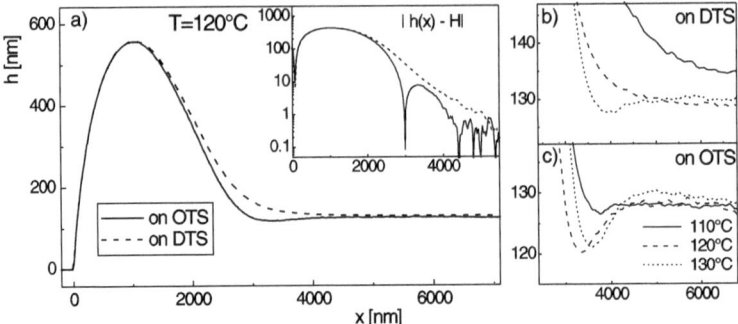

Fig. 4.13 Rim profiles of 130 nm PS films on DTS and OTS covered Si wafers, (**a**) at constant temperature. The *insert* shows a semilog plot of the height difference $|h(x) - h_f|$ with h_f as the initial flat film. The spiky features reveal damped oscillations in the film profile. Figures (**b**) and (**c**) show the decaying part of the profile for both surfaces, at three different temperatures. Reprinted with permission from Fetzer et al. (2005). Copyright by the American Physical Society

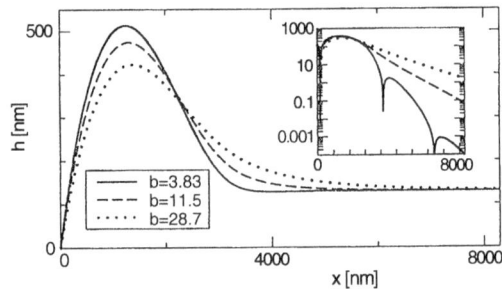

Fig. 4.14 Numerically calculated rim profiles for different slip lengths, non-dimensionalized by the height of the flat film. As in Fig. 4.13, the semilog-plot inset reveals oscillatory or non-oscillatory behaviour in the film profile decaying towards the film. Reprinted with permission from Fetzer et al. (2005). Copyright by the American Physical Society

and the thin-film equation yields, dropping all inertial terms and the interface potential contribution

$$4\beta\dot{s}\partial_\xi^2\varphi + \beta\partial_\xi^3\varphi - \dot{s}\varphi = 0. \tag{4.137}$$

We now seek for solutions of exponential type, $\phi = e^{\eta\xi}$, which yields the cubic equation

$$\beta\eta^3 + 4\beta\dot{s}\eta^2 - \dot{s} = 0. \tag{4.138}$$

From this cubic we then either find one positive real root, or either two negative or two complex conjugate roots. The first root is unphysical, so the relevant solutions are the real and complex conjugates whereby a change between two solutions occurs, as can be computed from the discriminant, at a critical slippage value $\dot{s}_c$ given by

$$\dot{s}_c = \frac{3}{16}\sqrt{\frac{3}{\beta}}. \tag{4.139}$$

This predicted change could indeed be induced experimentally by choosing adequate substrates. This was achieved in a series of papers by Fetzer et al. (2005, 2006, 2007a, 2007b), Bäumchen et al. (2009), Bäumchen and Jacobs (2010) where Si-wafer substrates were coated with either a monolayer of octadecyltrichlorosilane (OTS) or with dodecyltricholorosilane (DTS) (see Appendix A). Exemplary results are shown in Figs. 4.13 and 4.14.

4.7.5 Rim Instabilities

In this section we discuss the instabilities of the dewetting rim—we have seen this phenomenon on the figure at the beginning of the book, see also Fig. 4.15.

In the course of holes opening in the film, the rim continues to accumulate material: the rim thickens. It then runs into a well-known instability: the Rayleigh instability of a liquid column. Essentially two aspects render the instability of a dewetting rim is however different from that of a liquid column: firstly, the presence of a surface, and secondly, the fact that the rim is connected to a receding film and hence the whole rim is translated.

The first problem can be addressed by looking at the stability of a cylindrical liquid column on a substrate; this geometry is commonly called the 'ridge'-geometry, see Fig. 4.16. It has been discussed in detail by Sekimoto et al. (1987). Excitations of the ridge can be one of two types: the so-called 'zig-zag'-modes in which deformations of the two contact-lines of the ridge are in phase, and the 'varicose' or 'peristaltic' modes, in which they are out of phase. In the latter case, the width of the ridge is modulated in transverse direction, and hence in Laplace pressure. This pressure variation ultimately leads to a break-up of the ridge into droplets and the system reaches a state with lower surface area. If one denotes by $u_{\mathbf{q}}$ the outward displacement of the two contact lines the elastic energy per unit length for a varicose mode of wavevector $\mathbf{q}$ is given by Sekimoto et al. (1987), Redon et al. (1991), Brochard-Wyart and Redon (1992):

$$F_q = \sigma_{lv}\theta_e^2 q u_q^2 \left[-\frac{2}{wq} + \tanh\left(\frac{wq}{2}\right) \right] \tag{4.140}$$

whereby a limit $\kappa w \ll 1$ is assumed, where κ is the capillary length and w the width of the ridge: we see that in this consideration, the discussion was concerned with macroscopic ridges. But this limit also covers the case we are interested in, since the microscopic forces do not play a relevant role at the length scales of interest since the ridge, when it turns into the hydrodynamic instability, is already macroscopic (but not yet influenced by gravity). From Eq. (4.140) one finds that for $q < q_c$, i.e. for $q < 2.408/w$, the ridge is unstable.

Fig. 4.15 Time evolution of an opening hole in a PS film on DTS at 120 °C. Courtesy: Karin Jacobs

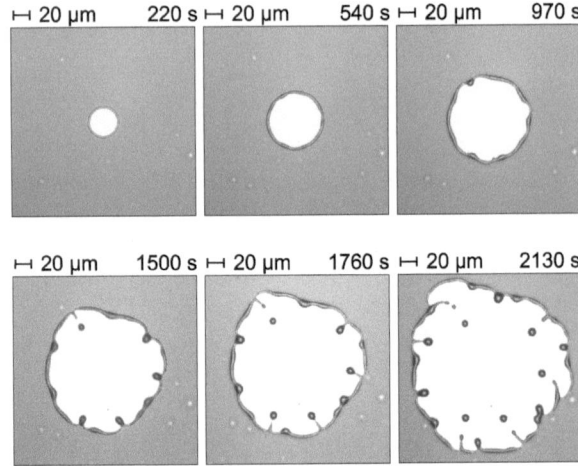

Fig. 4.16 Sketch of the liquid ridge geometry. *Arrows* point to the two contact lines of the ridge

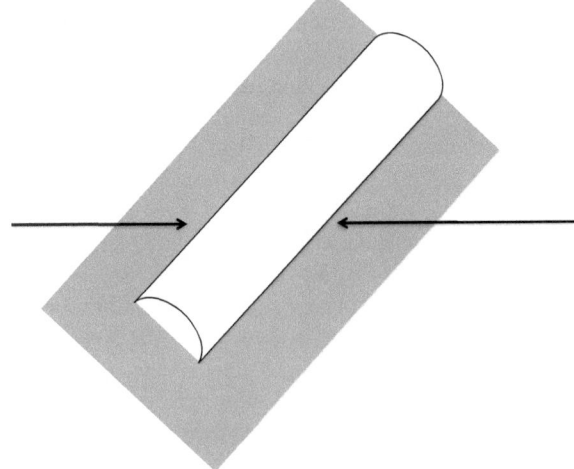

As for the evolution of the hole radius, analytic results based directly on the thin-film equations are difficult to obtain; there are numerical computations and some results from asymptotic analysis, see Münch and Wagner (2005), King et al. (2006), Münch and Wagner (2011). We here present a variant of a simplified calculation presented already quite some time ago. The relaxation spectrum of two coupled lines has been discussed in an approximate manner in Brochard-Wyart and Redon (1992).

Starting point is the energy balance

$$\dot{F}_q + T\dot{S} = 0 \tag{4.141}$$

where $T\dot{S}$ is viscous dissipation. In addition, one has to take into account the flow along the rim induced by the displacement of the two contact lines. The mean ve-

Fig. 4.17 Dewetting of a
linear ridge under strong slip.
Courtesy: Karin Jacobs

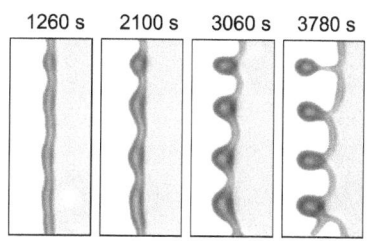

1260 s 2100 s 3060 s 3780 s

locity v_y is related to the displacement $\dot{u}$ by the incompressibility condition of the flow, $\nabla \cdot \mathbf{v} = 0$, which yields for the ridge geometry

$$\frac{2\dot{u}}{w} + qv_y = 0. \tag{4.142}$$

We can now derive the growth law of the ridge instability. For a single contact line, the dissipation is given by

$$T\dot{S} = -\frac{3\ell}{\theta}\eta\dot{u}^2, \tag{4.143}$$

hence for two we have a factor of 2. If $qw \ll 1$, the velocity in transverse direction is dominant, and one can get an approximate law

$$T\dot{S} = -\int dxdz\left(\frac{dv_y}{dz}\right)^2 \approx -\frac{\eta}{\theta}\frac{\dot{u}^2}{q^2w^2}, \tag{4.144}$$

and one interpolates

$$T\dot{S} = -\frac{6\ell}{\theta}\eta\left(1 + \frac{4}{q^2w^2}\right)\dot{u}^2, \tag{4.145}$$

whereby ℓ is a logarithmic factor related to the singularity of flow at the contact line; here it can be taken as a constant. Supposing further that the relaxation of u_q is exponential according to

$$\dot{u}_q = -(1/\tau_q)u_q \tag{4.146}$$

we obtain by putting everything together the expression

$$\frac{1}{\tau_q} = \frac{\sigma_{lv}}{3\ell\eta}\theta^3\frac{q}{1+4/(qw)^2}\left[-\frac{2}{wq} + \tanh\left(\frac{wq}{2}\right)\right]. \tag{4.147}$$

One finds that below a characteristic wavevector $q < q_c$, $1/\tau_q < 0$ and capillary excitations grow therefore exponentially. The risetime of the fastest mode is found to behave like $q_mw \approx 3/2$, hence $\lambda_m \approx 4.18w$.

The foregoing discussion concerns the case of a symmetric rim as it arises in a no- or weak-slip situation. In the case of strong slip, the rim is asymmetric (see before) and this asymmetry in fact becomes more pronounced in the course of the instability, see Fig. 4.17. A major difference between the weak-slip and strong-slip situation is the rise time of the instability: for strong slip, the onset of the instability is reached much faster. Replacing in this case ℓ by a slip-dependent factor

$$\ell = \ln(w/b) \tag{4.148}$$

as in Brochard-Wyart et al. (1994a) allows to rationalize this behaviour.

Chapter 5
Viscoelastic Thin Films

In this chapter we turn to viscoelastic effects in thin polymer films. We will see that this amounts, to a certain extent, to open up Pandora's box: life is much more complex here, and things are at present much less clear-cut than in the previously discussed viscous case.

We have seen before that we could describe thin-film behaviour on no- or weak-slip substrates very well with a simple Newton fluid model, and even better if we correct for film fluctuations. Also the strong-slip case could be treated mathematically and, by a proper tuning of substrate properties, compared to experiment. The general difficulty, however, is that, ultimately microscopically, slip and 'bulk' viscoelastic properties of thin films are not entirely independent from each other.

It is easy to get a qualitative idea why this should happen. (In fact, it should rather be astonishing that a mere viscous description works at all for chain-type molecules like polymers.) If we call ℓ_P some characteristic measure of the (dynamic) polymer length, we may assure the viscous regime if we have a separation of length scales

$$\ell_P \ll h_f \tag{5.1}$$

if we take h_f as the film thickness. If we now enlarge the polymer length relative to its confinement in the film, either by going to longer chains or to thinner films (or both), we are bound to leave a purely viscous regime. Dynamic effects of the polymers themselves will then enter into the story. In addition, we have to debate about the thermodynamic state of the polymer film. When we talk about polymeric thin films, we must first have created a *solid* film on the substrate (see Appendix A) which then heated to enter the fluid state. It is easy to imagine that the dewetting process and hence the polymer flow will depend on the state of 'solidity' of the film.

In this chapter we will first push ahead with the theoretical description of thin films in which viscoelastic properties arise and for this we will firmly stay on the side of 'macroscopic' hydrodynamics (Blossey 2008). Afterwards, we will revisit the problem from a more (but not entirely) microscopic perspective.

We first have to clarify what we understand under viscoelastic flow.

R. Blossey, *Thin Liquid Films*, Theoretical and Mathematical Physics,
DOI 10.1007/978-94-007-4455-4_5, © Springer Science+Business Media Dordrecht 2012

5.1 Viscoelastic Flow

5.1.1 Rate of Strain Tensor

In Chap. 4 we based our discussion of the dynamics of thin films on the purely viscous case in which we could write the *extra-stress tensor* is the simple form

$$\widehat{\tau} = \eta \nabla \mathbf{v} \tag{5.2}$$

with viscosity η. We now have to go beyond this simple behaviour, and this requires a more detailed discussion of the mathematical description of continuous media. In essence, we need to revisit the notions of *stress* and *strain*.

The notion of the stress tensor appeared already in the definition of the momentum equation, i.e., the Navier-Stokes equation we applied to the thin film geometry. So we now define what we understand under *strain*.

The application of a stress (a force) to a body can lead to its deformation. Strain is the proper measure, as it is defined as the relative distance between points of the body before and after the deformation. We therefore write the relationship of a spatio-temporal mass element $x(t)$ with such an element $x'(t')$ as

$$d\mathbf{x}' = \widehat{F}(x,t,t')d\mathbf{x}, \qquad \widehat{F}_{ij} = \frac{\partial x_i'}{\partial x_j}. \tag{5.3}$$

The *deformation gradient tensor* $\widehat{F}$ is not necessarily symmetric. We can write

$$(d\mathbf{x}')^2 = (d\mathbf{x}')^T \cdot (d\mathbf{x}') = (\widehat{F}d\mathbf{x})^T \cdot (\widehat{F}d\mathbf{x}) = (d\mathbf{x})^T (\widehat{F}^T \cdot \widehat{F})d\mathbf{x} \tag{5.4}$$

where the superscript T denotes the transpose. We now define the *strain tensor* as

$$\widehat{\gamma} \equiv \widehat{F}^T \cdot \widehat{F}. \tag{5.5}$$

This tensor is also sometimes called the *right-relative Cauchy-Green tensor*.

We can consider time derivatives of the strain tensor $\widehat{\gamma}$. Since we have

$$\widehat{\gamma}_{ij}(t') = \widehat{F}_{ki}\widehat{F}_{kj} = \frac{\partial x_k'}{\partial x_i}\frac{\partial x_k'}{\partial x_j} \tag{5.6}$$

the primed coordinates x_k' depend on t', but the unprimed coordinates do not, and therefore time differentiation yields

$$\frac{d\widehat{\gamma}_{ij}}{dt'} = \frac{\partial x_k'}{\partial x_i}\frac{\partial u_k}{\partial x_j} + \frac{\partial u_k}{\partial x_i}\frac{\partial x_k'}{\partial x_j} \tag{5.7}$$

which for $t = t'$, $x_k = x_k'$ leads to

$$\frac{d\widehat{\gamma}_{ij}}{dt'}\bigg|_{t'=t} = \widehat{\delta}_{ki}\frac{\partial u_k}{\partial x_j} + \frac{\partial u_k}{\partial x_i}\widehat{\delta}_{kj} = \frac{\partial u_i}{\partial x_j} + \frac{\partial u_j}{\partial x_i} \equiv 2\dot{\widehat{\gamma}}_{ij} \tag{5.8}$$

and we have obtained the *rate of strain tensor*

$$\dot{\hat{\gamma}} = \frac{1}{2}\left[\nabla \mathbf{u} + (\nabla \mathbf{u})^T\right]. \tag{5.9}$$

5.1.2 The Jeffreys Model

In order to close the relation we need to relate the *extra-stress tensor* $\hat{\tau}$ to the *rate of strain tensor* $\hat{\gamma}$.

The relation between these two tensors is not unique: it is a material-dependent property, and consequently various degrees of complexity can be built into this relation. There are two ways of constructing it: either by the way of phenomenology, or from a microscopic approach involving molecular kinetic models.

The latter path has, for polymeric thin films, not yet been properly attempted (but well for bulk polymers, see the final section of this chapter). Most models discussed in the literature have, one way or the other, been variants of one of the simplest models for viscoelastic media, the Jeffreys model. It has enjoyed prime interest certainly because it introduces a minimal number of additional parameters; which is, in some sense, required in order not to end up with ambiguous parametrization problems.

In order to understand this model, we must find a suitable generalization of the expression of the extra-stress tensor from a viscous to a viscoelastic fluid. The former is given by, employing the rate of strain tensor, by the expression

$$\hat{\tau} = \frac{\eta}{2}\left[\nabla \mathbf{u} + (\nabla \mathbf{u})^T\right] = \eta\dot{\hat{\gamma}}. \tag{5.10}$$

A suitable generalization is to introduce a *memory kernel* or *relaxation function* in the form (Böhme 2000)

$$\hat{\tau} = \int_0^\infty ds\, G(s)\dot{\hat{\gamma}}(t-s), \tag{5.11}$$

such that for constant $\dot{\hat{\gamma}}$,

$$\int_0^\infty ds\, G(s) \equiv \eta \tag{5.12}$$

holds.

If a system responds to the application of a stress with an exponential relaxation, we have

$$G(s) = G(0)\exp(-s/\lambda)H(s) \tag{5.13}$$

where $H(s)$ is the Heaviside function, $H(s) = 0$ for $s < 0$, $H(s) = 1$ for $s \geq 0$. Our simple model depends on two parameters, the amplitude $G(0)$ and the relaxation

Fig. 5.1 Equivalent circuit diagram of the Maxwell and the Jeffreys models. The explanation of the symbols can be found in the text

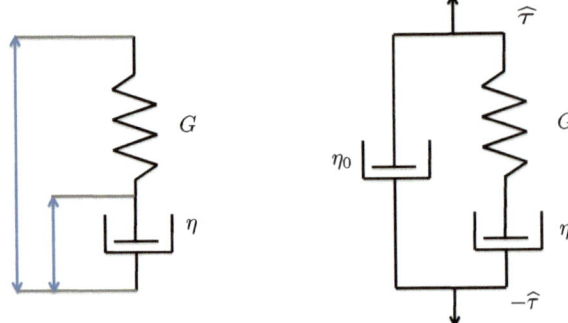

time λ. The presence of this relaxation behaviour is characteristic of elastic behaviour of the fluid. This follows from the differentiation of Eq. (5.11) with respect to time, employing Eq. (5.13), which yields

$$\dot{\widehat{\tau}}(t) = -G(0) \int_0^\infty ds\, e^{-s/\lambda} \frac{d\dot{\widehat{\gamma}}}{ds}. \tag{5.14}$$

Following a partial integration, we end up with the differential relation

$$\dot{\widehat{\tau}}(t) + G(0)\dot{\widehat{\gamma}}(t) = -\frac{1}{\lambda}\widehat{\tau}(t). \tag{5.15}$$

We can rewrite this in the form

$$\widehat{\tau}(t) + \lambda\dot{\widehat{\tau}}(t) = \eta\dot{\widehat{\gamma}}(t), \tag{5.16}$$

where we denote $\eta \equiv G(0)\lambda$.

Equation (5.16), corresponds to the simplest viscoelastic model, the *Maxwell model*. It contains a purely viscous damping given by η and an elastic spring with modulus $G(0) = \eta/\lambda$. We can also introduce, alongside the total strain $\widehat{\gamma}$, the strain associated with the damping term, and we can relate them to the total stress via

$$\widehat{\tau} = \eta\dot{\widehat{\gamma}}_1 = G(0)(\widehat{\gamma} - \widehat{\gamma}_1) \tag{5.17}$$

which is equivalent to Eq. (5.16). We can represent the Maxwell model by an *equivalent circuit diagram* in an electrotechnical analogy which is shown in Fig. 5.1. In this diagram, elastic behaviour is represented by a spring, while viscous behaviour is depicted by a damper.

More complex rheological models now simply follow by the addition of more springs or dampers. The *Jeffreys model* is simply the next in an ascending line of more complex models, specifically here in which one additional damping is added, see Fig. 5.1. The additional viscous term which acts in parallel to the Maxwell system. One thus has

$$\widehat{\tau} = \eta_0\dot{\widehat{\gamma}} + \widehat{\tau}_1 \tag{5.18}$$

where $\widehat{\tau}_1$ is the stress in the right branch of the diagram in Fig. 5.1. For the latter, we have the Maxwell relation

$$\widehat{\tau}_1(t) + \lambda_1 \dot{\widehat{\tau}}_1(t) = \eta_1 \dot{\widehat{\gamma}}(t). \tag{5.19}$$

Eliminating $\widehat{\tau}_1$ from the two equations we obtain

$$\widehat{\tau}(t) + \lambda_1 \dot{\widehat{\tau}}(t) = (\eta_0 + \eta_1)\dot{\widehat{\gamma}}(t) + \lambda_1 \eta_0 \ddot{\widehat{\gamma}}(t), \tag{5.20}$$

in which now a *deformation acceleration* appears. We can define

$$\eta \equiv \eta_0 + \eta_1, \qquad \lambda_2 \equiv \frac{\eta_0}{\eta_0 + \eta_1}\lambda_1. \tag{5.21}$$

One notes that, in fact, $\lambda_2 > \lambda_1$, and $\eta \approx \eta_1$.

In this subsection we employ a basic version of the Jeffreys model which carries variations in time. However, also spatial dependencies can be considered, leading to what is called *convective terms*. We will come back this generalization later in the book.[1]

5.2 Lubrication Approximation III

5.2.1 Weak Slip

In this subsection we return to the hydrodynamic problem, first for weakly slipping viscoelastic liquids of Jeffreys type. Here, as in the two following subsections, we will derive lubrication models which cover again the range from weak- to strong slip, but now in the presence of viscoelasticity. The method employed is as before the asymptotic analysis which was presented in detail in the previous chapter. We will therefore not go again into every detail of these computations which, when needed, can be found in the original literature.

In Rauscher et al. (2005) the thin-film equation for a viscoelastic film was derived under the assumption of weak slip. The model equations thus are first mass conservation as in the incompressible case

$$\nabla \cdot \mathbf{u} = 0 \tag{5.22}$$

for the velocity field $\mathbf{u} = (u_x, u_y, u_z)$. *Momentum conservation* is given by[2]

$$\varrho \frac{d\mathbf{u}}{dt} = -\nabla p_R + \nabla \cdot \tau \tag{5.23}$$

where now $d/dt = \partial_t + \mathbf{u} \cdot \nabla$ is the total (or material) derivative.

---

[1] Note that the Jeffreys model is used in several parametrizations in the literature. Upon multiplying Eq. (5.20) with the elastic modulus G we can find the form $G\widehat{\tau} + \eta_1\dot{\widehat{\tau}} = G\eta_1\dot{\widehat{\gamma}} + \eta_0\eta_1\ddot{\widehat{\gamma}}$ which is used by Vilmin and Raphaël (2006).

[2] We omit the ⌢-symbol in the following to simplify notation.

In a Newtonian liquid, the traceless part of the stress tensor τ is proportional to the strain rate, i.e. the gradient of the velocity field,[3]

$$\dot{\gamma}_{ij} = \partial_i u_j + \partial_j u_i. \tag{5.24}$$

In a purely elastic medium, the stress is proportional to the strain. In a viscoelastic medium, one has to model the dependence $\tau(\dot{\gamma})$ which interpolates between the elastic and the viscous regime. In the Jeffreys model introduced before one assumes the linear relation between stress and strain and their time derivatives of the form

$$\tau + \lambda_1 \partial_t \tau = \eta(\dot{\gamma} + \lambda_2 \partial_t \dot{\gamma}) \tag{5.25}$$

which contains the two relaxation time constants λ_1 and λ_2 as well as the shear viscosity η.

In the context of the thin polymer films we interpret the viscoelastic response of the liquid in the following way. There are now two viscous regimes, an 'early' and a 'late' regime, into which an elastic response is interspersed. This means that for short times, the response of the liquid to a shear is first viscous until elastic effects start to intervene. After the relaxation of elastic stresses, the film enters the previously discussed viscous flow regime.

We turn to the lubrication calculation. The parametrization of the thin film is done identically to the lubrication calculation in the purely viscous case. The difference between the two cases arises in the components of the strain-rate tensor. If we assume that corresponding components of the stress and strain rate tensor are of the same order, we obtain the following scalings

$$\mathbf{r}_{\parallel} = L \mathbf{r}_{\parallel}^*, \quad (z, h, b) = H(z^*, h^*, b^*), \tag{5.26}$$

$$\mathbf{u}_{\parallel} = U \mathbf{u}_{\parallel}^*, \quad (t, \lambda_1, \lambda_2) = T(t^*, \lambda_1^*, \lambda_2^*), \tag{5.27}$$

$$u_z = \varepsilon U u_z^*, \quad (p, V, p_R) = P(p^*, V^*, p_R^*), \tag{5.28}$$

$$\sigma = \frac{U\eta}{\varepsilon^3}\sigma^* \tag{5.29}$$

and hence we have for the extra-stress tensor

$$\begin{pmatrix} \tau_{xx} & \tau_{xy} & \tau_{xz} \\ \tau_{yx} & \tau_{yy} & \tau_{yz} \\ \tau_{zx} & \tau_{zy} & \tau_{zz} \end{pmatrix} = \frac{\eta}{T} \begin{pmatrix} \tau_{xx}^* & \tau_{xy}^* & \dfrac{\tau_{xz}^*}{\varepsilon} \\ \tau_{yx}^* & \tau_{yy}^* & \dfrac{\tau_{yz}^*}{\varepsilon} \\ \dfrac{\tau_{zx}^*}{\varepsilon} & \dfrac{\tau_{zy}^*}{\varepsilon} & \tau_{zz}^* \end{pmatrix} \tag{5.30}$$

---

[3]One notes a difference of a factor of $1/2$ between this equation and equation (5.9). This is convention-dependent. We use the convention used in Rauscher et al. (2005). In comparing both cases, it suffices to assume that the additional factor has been absorbed in the definition of the viscosity η, see Eq. (5.25) below.

where the * denotes dimensionless quantities. We note that this scaling prescription, although physically motivated, is not the only one used in the literature, see e.g. Khayat (2001), Zhang et al. (2002).

With this scaling we obtain the dimensionless dynamic equations in the following form:

$$\varepsilon^2 Re \frac{du_i}{dt} = \varepsilon^2 (\partial_x \tau_{xi} + \partial_y \tau_{yi}) + \partial_z \tau_{zi} - \partial_i p_R \tag{5.31}$$

for $i = x, y$. For the normal component we have

$$\varepsilon^4 Re \frac{du_z}{dt} = \varepsilon^2 (\partial_x \tau_{xz} + \partial_y \tau_{yz} + \partial_z \tau_{zz}) - \partial_z p_R \tag{5.32}$$

again with the Reynolds number $Re = \varrho U L / \eta$.

The non-dimensionalized Jeffreys model is given by ($i = x, y, z$),

$$\tau_{ii} + \lambda_1 \partial_t \tau_{ii} = 2(\partial_i u_i + \lambda_2 \partial_t \partial_i u_i) \tag{5.33}$$

$$\tau_{xy} + \lambda_1 \partial_t \tau_{xy} = \dot{\gamma}_{xy} + \lambda_2 \partial_t \dot{\gamma}_{xy} \tag{5.34}$$

with $\dot{\gamma}_{xy} = \partial_x u_y + \partial_y u_x$, and

$$\tau_{iz} + \lambda_1 \partial_t \tau_{iz} = \partial_z u_i + \lambda_2 \partial_t \partial_z u_i + (\partial_i u_z + \lambda_2 \partial_t \partial_i u_z)\varepsilon^2 \tag{5.35}$$

with $i = x, y$.

The boundary conditions for this problem apply in the same form as they were formulated in Chap. 4, since there we used already a general form for the extra-stress tensor, hence no additional steps are needed in the derivation.

We can now move on to the thin-film equation. For the parallel and normal momentum equations we have

$$\partial_z \tau_{zi} = \partial_i p, \qquad \partial_z p = 0, \tag{5.36}$$

for $i = x, y$. The equations of the Jeffreys model either do not contain ε, or second-order terms which can be dropped and the lowest-order be read off immediately. The boundary conditions at the film surface $z = h(x, y, t)$ are to leading order given by

$$p_R = -\nabla_\parallel^2 h + \phi'(h) \tag{5.37}$$

$$\partial_x h \tau_{xz} + \partial_y h \tau_{yz} = 0 \tag{5.38}$$

$$\partial_x h \tau_{yz} - \partial_y h \tau_{xz} = 0 \tag{5.39}$$

For $\nabla_\parallel h \neq 0$, we have $\tau_{xz} = \tau_{yz} = 0$. Since the flow field $\mathbf{u}$, the pressure p and therefore h do only depend on τ_{xz} and τ_{yz}, we have a closed system of equations for the variables $\mathbf{u}$, p, h, τ_{xz} and τ_{yz}.

As before we can at this point now pass on to the integration of the equations. According to the normal component of the momentum equation, p_R is independent of z. Integrating the parallel components of the momentum equation with respect to z from z to h yields

$$\tau_{iz} = (z - h)\partial_i p_R. \tag{5.40}$$

Substitution of this result into the constitutive equation yields

$$(1 + \lambda_2 \partial_t)\partial_z u_i = (1 + \lambda_1 \partial_t)\left[(z - h)\partial_i p_R\right] \tag{5.41}$$

which can be integrated from 0 to z, making use of the boundary condition for u_i and the value of τ_{iz} at $z = 0$,

$$(1 + \lambda_2 \partial_t)(u_i + bh\partial_i p_R) = (1 + \lambda_1 \partial_t)\left[\left(\frac{z^2}{2} - hz\right)\partial_i p_R\right]. \tag{5.42}$$

Integrating this result now once more from $z = 0$ to $z = h(x, y, t)$ we obtain a somewhat lengthy expression

$$(1 + \lambda_2 \partial_t)\left(\int_0^h dz u_i + bh^2 \partial_i p_R\right) - \lambda_2 \partial_t h(u_i|_{z=h} + bh\partial_i p_R)$$

$$= -(1 + \lambda_1 \partial_t)\left(\frac{h^3}{3}\partial_i p_R\right) + \lambda_1 \frac{h^2}{2}\partial_t h \partial_i p_R. \tag{5.43}$$

With the use of the kinematic condition (4.22) we obtain the expression

$$\partial_t h + \lambda_2\left[\partial_t^2 h + \nabla_\| \cdot (\mathbf{u}_\||_{z=h}\partial_t h)\right]$$

$$= \nabla_\| \cdot \left\{\left[(1 + \lambda_1 \partial_t)\frac{h^3}{3} + (1 + \lambda_2 \partial_t)bh^2\right.\right.$$

$$\left.\left. - \left(\frac{h^2}{2}\lambda_1 + bh\lambda_2\right)\partial_t h\right]\nabla_\|(p_R)\right\} \tag{5.44}$$

In order to close the equation, we have to find an expression for $\mathbf{u}|_{z=h}$ in terms of $h(x, y, t)$. Observing that Eq. (5.42) can be written as an ordinary differential equation in time

$$u_i + \lambda_2 \partial_t u_i = g_i \tag{5.45}$$

where

$$g_i \equiv -(1 + \lambda_2 \partial_t)bh\partial_i p_R + (1 + \lambda_1 \partial_t)\left[\left(\frac{z^2}{2} - hz\right)\partial_i p_R\right] \tag{5.46}$$

we can represent the solution in the form

$$u_i = \frac{1}{\lambda_2}\int_{-\infty}^t dt' \exp\left(-\frac{t - t'}{\lambda_2}\right)g_i\left(x, y, z, t'\right) \equiv \frac{1}{\lambda_2}\mathcal{L}[g_i]. \tag{5.47}$$

This result can now be simplified by partial integration to yield

$$\lambda_2 \mathbf{u}_\| |_{z=h} = -\left(\lambda_1 \frac{h^2}{2} + \lambda_2 bh\right) \nabla_\| p_R + (\lambda_2 - \lambda_1)\left(\frac{h^2}{2}\mathbf{Q}_\| - h\mathbf{R}_\|\right), \tag{5.48}$$

where we have

$$\mathbf{Q}_\| = \frac{1}{\lambda_2}\mathcal{L}[\nabla_\| p_R], \qquad \mathbf{R}_\| = \frac{1}{\lambda_2}\mathcal{L}[h\nabla_\| p_R], \tag{5.49}$$

or, equivalently,

$$\mathbf{Q}_\| + \lambda_2 \partial_t \mathbf{Q}_\| = \nabla_\| p_R, \qquad \mathbf{R}_\| + \lambda_2 \partial_t \mathbf{R}_\| = h\nabla_\| p_R. \tag{5.50}$$

Putting it altogether, we end up with the following expression for the lubrication equation for h, which needs to be considered together with the equations for $\mathbf{Q}_\|$ and $\mathbf{R}_\|$, (5.49) or (5.50):

$$(1 + \lambda_2 \partial_t)\partial_t h + (\lambda_2 - \lambda_1)\nabla_\| \cdot \left[\left(\frac{h^2}{2}\mathbf{Q}_\| - h\mathbf{R}_\|\right)\partial_t h\right]$$

$$= \nabla_\| \cdot \left\{\left[(1 + \lambda_1 \partial_t)\frac{h^3}{3} + (1 + \lambda_2 \partial_t)bh^2\right]\nabla_\| p_R\right\} \tag{5.51}$$

After this little *tour de force* in deriving the thin-film equation for a Jeffreys fluid in the lubrication limit, we can pause a little and observe some of its properties than can be immediately read off.

Firstly, we note that the dependence of the Jeffreys model on higher-order derivatives of the stress and strain tensors is reflected by a second-order derivative of the film height with respect to time. Mixing of space- and time derivatives makes the equation considerably more complex than for the purely viscous case.

Secondly, some limiting cases emerge. For $\lambda_2 \to 0$, the equation collapses to a single equation which corresponds to a Maxwell model with only one stress tensor contribution. In the case of $\lambda_1 = \lambda_2$, we recover the thin-film equation of a Newtonian liquid, multiplied by a factor of $(1 + \lambda_1 \partial_t)$.

5.2.2 Strong Slip

In this section we complete the description by deriving the analogue for viscoelastic liquids of the strong-slip equation for viscous liquids. The derivation turns out to be much simpler than for the weak-slip case (Blossey et al. 2006).

As for the purely viscous liquid, the main difference between the weak-slip and the strong-slip case resides in the scaling. Taking again, as we did for the viscous strong-slip case, the basic length scalings as

$$z = Hz^*, \qquad (x, y) = \left(Lx^*, Ly^*\right), \qquad b = Hb^*, \tag{5.52}$$

and assume the usual scaling ratio $\varepsilon \equiv H/L \ll 1$ and time $T = L/U$. The stress tensor scales as

$$\tau_{ij} = \frac{\eta}{T}\tau_{ij}^*, \quad i = j \tag{5.53}$$

for the diagonal components, while for $i \neq j$ we have

$$\tau_{ij} = \frac{\eta}{\varepsilon T}\tau_{ij}^*, \quad i = j \tag{5.54}$$

The scalings thus are as in the viscous case of strong slip. Also, for the derivation of the thin-film equations, our starting point is the *ansatz*

$$(u, w, h, p_R, \tau_{ij}) = (u_0, w_0, h_0, p_{R0}, \tau_{ij0}) + \varepsilon^2(u_1, w_1, h_1, p_{R1}, \tau_{ij1}). \tag{5.55}$$

To leading order one has

$$\tau_{xz0} = 0, \tag{5.56}$$

$$(1 + \lambda_2\partial_t)\partial_z u_0 = 0 \tag{5.57}$$

with the solution

$$\partial_z w_0 = c(w, z)\exp(-\lambda_2 t). \tag{5.58}$$

We select the solution $c \equiv 0$ since any other solution would correspond to a strong prestressing of the film at times $t \to -\infty$ (a point to which we return in the discussion of experiments). Therefore, $w_0 = f(x, t)$, and from mass conservation we have $\partial_x f = -\partial_z w_0$, hence $w_0 = -z\partial_x f$. Thus

$$(1 + \lambda_1\partial_t)\tau_{zz0} = -(1 + \lambda_2\partial_t)\partial_x f \tag{5.59}$$

which in integrated from reads as

$$\tau_{zz0} = -\frac{2}{\lambda_1}\int_{-\infty}^{0} dt' e^{(t-t')/\lambda_1}(1 + \lambda_2\partial_t)\partial_x f = -\tau_{xx0}. \tag{5.60}$$

To solve for $f(x, t)$ we need to make use of the next order which gives

$$Re^*(\partial_t + f\partial_x)f = \partial_x\tau_{xx0} + \partial_z\tau_{xz1} = -\partial_x p_{R0}, \tag{5.61}$$

where $p_{R0} = -\partial_{xx}h_0 - \tau_{zz0}$. This can be written as

$$Re^*(\partial_t + f\partial_x)f = \partial_z\tau_{xz1} + \partial_{xxx}h_0 + \frac{4}{\lambda_1}\partial_x\int_{-\infty}^{t} dt' e^{(t-t')/\lambda_1}(1 + \lambda_2\partial_t)\partial_x f. \tag{5.62}$$

From the boundary condition at the free surface we find to second order

$$((\tau_{xx0} - \tau_{zz0}) + \tau_{xz0}(\partial_x h_0))(\partial_x h_0) = \tau_{xz1} \tag{5.63}$$

and

$$\tau_{xz1} = -2(\partial_x h_0)\tau_{zz0}. \tag{5.64}$$

It remains to determine the second-order result from the boundary condition at the substrate. We have $\tau_{xz1} = f/\beta$ and can now integrate Eq. (5.62) with respect to z across the film from 0 to h_0 and obtain the following system of equations, where $q = -\tau_{zz0}/2$:

$$h Re^*(\partial_t u + u\partial_x u) = h\partial_x\left[\partial_x^2 h - \phi''(h)\right] + \partial_x(4hq) - \frac{u}{\beta}, \tag{5.65}$$

$$(1 + \lambda_1\partial_t)q = (1 + \lambda_2\partial_t)\partial_x u, \tag{5.66}$$

$$\partial_t h + \partial_x(hu) = 0. \tag{5.67}$$

We can conclude that the generalization of the strong-slip system of equations in the case of the Jeffreys model is very natural—which is quite a contrast to the weak-slip case, in which the resulting equation looks quite complicated in the viscoelastic case.

We will now see what can be learnt from the equations, and we do so but first studying the dispersion relations for a thin film.

5.2.3 Dispersion Relations

Does viscoelasticity have an influence on film rupture? This can be checked by looking at the dispersion relation, as before, for the case of spinodal dewetting.

For the weak-slip equation, one assumes

$$h = h_0 + \delta h_1, \qquad Q = \delta Q_1, \qquad R = \delta R_1 \tag{5.68}$$

with $0 < \delta \ll 1$, and

$$(h_1, Q_1, R_1) = (h_1, Q_1, R_1)e^{iqx+\omega t}. \tag{5.69}$$

The resulting dispersion relation $\omega(q)$ can then be expressed as

$$(1 + \lambda_2\omega)\omega = \omega_N(q)(1 + \Lambda\omega) \tag{5.70}$$

where

$$\omega_N(q) = M(h_0)\Omega(q) \tag{5.71}$$

is the dispersion relation of the Newtonian thin film. The parameter Λ is given by

$$\Lambda \equiv \lambda_2 + \frac{(\lambda_1 - \lambda_2)h_0^3}{h_0^3 + 3bh_0^2}. \tag{5.72}$$

Equation (5.70) has two branches of solutions, one of which is strictly negative and one which has the same sign and zeroes as $\omega_N(q)$. Since furthermore the wavevector q does not enter explicitly in the equation for $\omega(q)$, also the fastest mode is unaffected by viscoelastic relaxation. The instability is therefore the same as for a Newtonian film: in the weak-slip case, viscoelasticity has no (immediate) effect on film rupture.

For the strongly slipping film, we put

$$h = h_0 + \delta h_1, \qquad q = \delta q_1, \qquad u = \delta u_1 \tag{5.73}$$

and we are led to the dispersion relation

$$(1+\lambda_1\omega)\big(h_0 Re^*\omega + \beta^{-1}\big)\omega + 4h_0 q^2\omega(1+\lambda_2\omega) - h_0^2\Omega(q)(1+\lambda_1\omega) = 0. \tag{5.74}$$

Again, one finds a solution branch with the same zeroes as $\Omega(q)$—also here, the range of wave-vectors is unaffected by viscoelastic relaxation.

However, in contrast to the weak-slip case, the most unstable wavevector q_m is now modified. Considering for simplicity the case $Re^* = 0$, it fulfills the equation

$$4\beta h_0^3 q_m^4 + h_0^2\big(2q_M^2 + V''(h_0)\big)\frac{1+\lambda_1\beta h_0^2 q_m^4}{1+\lambda_2\beta h_0^2 q_m^4} = 0. \tag{5.75}$$

In the limit $\lambda_1, \lambda_2 \gg 1$, and $\beta \gg 1$ we find

$$q_m^2 = -\frac{\varrho}{4} + \sqrt{\frac{\varrho^2}{16} - \frac{V''(h_0)\varrho}{4}}, \qquad \varrho = \frac{\lambda_1}{\beta h_0 \lambda_2}. \tag{5.76}$$

We note that the result depends on the combination of systems lengths $\sim h_0\beta$, and that the most unstable mode diverges for $\beta \to \infty$, a result also found by Kargupta et al. (2004) already for the case of a Newtonian liquid. Hence slip does strongly affect the most unstable mode.

In the case of a viscoelastic film discussed here, there is the additional dependence on the relaxation parameters λ_1 an λ_2. But as we see in the expression

$$\varrho = \frac{\lambda_1}{\beta h_0 \lambda_2} \tag{5.77}$$

the parameters are so intertwined with the slip length that, at this level, slip and viscoelastic effects are difficult to disentangle. We therefore move on to the dewetting rims in viscoelastic thin films.

5.2.4 Dewetting Rims

In this subsection we look at whether for a weakly slipping viscoelastic film, the profile of a dewetting hole will display oscillations or whether viscoelasticity alone can bring about a monotonous rim profile. We will do this calculation in two dimensions.

Technically, the problem is identical to what we have discussed already in Chap. 4 for the viscous case. As before, we shift to a comoving frame, $\xi \equiv x - s(t)$, but now for the three functions

$$h(x,t) = h(\xi,t), \qquad Q(x,t) = Q(\xi,t), \qquad R(x,t) = R(\xi,t) \qquad (5.78)$$

where Q and R are the first components of $\mathbf{Q}_\|$ and $\mathbf{R}_\|$, respectively. This yields the following fairly unwieldy expression

$$\partial_t h - \dot{s}\partial_\xi h + \lambda_2\left(\partial_t^2 h - 2\dot{s}\partial_t\partial_\xi h + \dot{s}^2\partial_\xi^2 h - \ddot{s}\partial_\xi h\right)$$

$$+ (\lambda_2 - \lambda_1)\partial_\xi\left[(\partial_t h - \dot{s}\partial_\xi h)\left(\frac{h^2}{2}Q - hR\right)\right]$$

$$= \partial_\xi\left[-\left(\frac{h^3}{3} + bh^2\right)\partial_\xi^3 h - (\partial_t - \dot{s}\partial_\xi h)\left\{\left(\lambda_1\frac{h^3}{3} + \lambda_2 bh^2\right)\partial_\xi^3 h\right\}\right] \qquad (5.79)$$

together with

$$Q + \lambda_2\partial_t Q - \lambda_2\dot{s}\partial_\xi Q = -\partial_\xi^3 h \qquad (5.80)$$

and

$$R + \lambda_2\partial_t R - \lambda_2\dot{s}\partial_\xi R = -h\partial_\xi^3 h. \qquad (5.81)$$

We now perturb around the flat reference state with $h_0 = \text{const.}$, $Q = R = 0$, setting

$$h = h_0 + \delta\varphi, \qquad Q = \delta\psi_1, \qquad R = \delta\psi_2 \qquad (5.82)$$

and, in addition, assume a quasi-steady state in which the shape of the rim changes only slowly and the speed $\dot{s}$ is constant. Keeping only the lowest order terms in δ we find

$$-\dot{s}\partial_\xi\varphi + \lambda_2\dot{s}^2\partial_\xi^2\varphi + \left(\frac{h_0^3}{3} + bh_0^2\right)\partial_\xi^4\varphi - \dot{s}\left(\lambda_1\frac{h_0^3}{3} + \lambda_2 bh_0^2\right)\partial_\xi^5\varphi = 0 \qquad (5.83)$$

and

$$\psi_1 - \lambda_2\dot{s}\partial_\xi\psi_1 = -\partial_\xi^3\varphi \qquad (5.84)$$

$$\psi_2 - \lambda_2\dot{s}\partial_\xi\psi_2 = h_0\partial_\xi^3\varphi. \qquad (5.85)$$

Since Eq. (5.83) does not contain any contributions from ψ_1 and ψ_2, we can simply solve it by making the normal mode *ansatz* $\varphi = \exp(\omega\xi)$, requiring that the solutions decay for $\varphi \to 0$ since $h \to h_0$, $Q = 0$ and $R = 0$ as $\xi \to \infty$. Hence, the solutions must always have an ω with a negative real part.

However, we find that in the resulting growth equation

$$-\dot{s} + \lambda_2 \dot{s}\omega + \left(\frac{h_0^3}{3} + bh_0^2\right)\omega^3 - \dot{s}\left(\lambda_1 \frac{h_0^3}{3} + \lambda_2 bh_0^2\right)\omega^4 = 0 \qquad (5.86)$$

all coefficients are *positive* constants. We can therefore conclude from the form of the polynomial that normal modes with negative ω never occur. Consequently, the solutions have to have oscillatory behaviour, as in the case of a simple Newtonian fluid film. Viscoelasticity alone (at least of Jeffreys type) *cannot* bring about a monotone rim profile.

5.3 Beyond the Jeffreys Model

5.3.1 The Corotational Jeffreys Model

In this section we discuss a further extension of our viscoelastic model, its *corotational generalization*, as alluded to before (Münch et al. 2006).

$$\widehat{\tau} + \lambda_1 \frac{D_J \widehat{\tau}}{D_J t} = \mu\left(\widehat{\dot{\gamma}} + \lambda_2 \frac{D_J \widehat{\dot{\gamma}}}{D_J t}\right) \qquad (5.87)$$

where here

$$\frac{D_J \widehat{\Lambda}}{D_J t} = \frac{d\widehat{\Lambda}}{dt} + \frac{1}{2}(\widehat{\omega}\widehat{\Lambda} - \widehat{\Lambda}\widehat{\omega}) \qquad (5.88)$$

is the *Jaumann derivative*, with the rate of strain tensor $\widehat{\dot{\gamma}}$ and the *vorticity tensor*

$$\widehat{\omega} = \nabla\mathbf{u} - \nabla\mathbf{u}^T. \qquad (5.89)$$

Here, as before, $d/dt = \partial_t + \mathbf{u} \cdot \nabla$. This model is one of a class of generalizations of viscoelastic flows, involving generalized time-derivatives, reflecting the allowed transformation properties of the involved tensorial fields. The simplest case consists in a *convected derivative*, which corresponds to a coordinate base is transported along with the flow (*upper convected* or *contravariant convected derivative*). If the unit basis is chosen normal to coordinate planes, one obtains the *lower convected* or *covariant convected derivative*. Both bases translate, rotate and deform with the material during its flow. Our choice of time derivative corresponds to a linear combination of these bases, in which the resulting basis then rotates with the flow, but does not deform. For a much more detailed discussion of the different viscoelastic models, we refer the reader to Byron Bird et al. (1987a, 1987b).

We again restrict the calculations to the 2D-case and use, as before, the velocity field components (u, w) and employ the strong-slip scaling. In this scaling, the friction between the liquid and the substrate is too weak to maintain a non-zero (xz)-shear stress to lowest order, and lateral pressure gradients are balanced by the

(xx)-component of the stress tensor. The dimensional stress tensor reads, in terms of the non-dimensional components, as

$$\frac{\mu}{T} \begin{pmatrix} \tau^{xx} & \frac{1}{\varepsilon}\tau^{xz} \\ \frac{1}{\varepsilon}\tau^{xz} & \tau^{zz} \end{pmatrix}$$

The dimensionless corotational Jeffreys model can then be written as

$$\left(1 + \lambda_1 \frac{d}{dt}\right)\tau^{xx} - \lambda_1\left(\frac{1}{\varepsilon^2}\partial_z u - \partial_x w\right)\tau^{xz}$$

$$= 2\mu\left(1 + \lambda_2\frac{d}{dt}\right)\partial_x u - \lambda_2\mu\left(\frac{1}{\varepsilon^2}(\partial_z u)^2 - \varepsilon^2(\partial_x w)^2\right), \tag{5.90}$$

$$\left(1 + \lambda_1 \frac{d}{dt}\right)\tau^{zz} + \lambda_1\left(\frac{1}{\varepsilon^2}\partial_z u - \partial_x w\right)\tau^{xz}$$

$$= 2\mu\left(1 + \lambda_2\frac{d}{dt}\right)\partial_z w + \lambda_2\mu\left(\frac{1}{\varepsilon^2}(\partial_z u)^2 - \varepsilon^2(\partial_x w)^2\right), \tag{5.91}$$

$$\left(1 + \lambda_1 \frac{d}{dt}\right)\tau^{xx} + \frac{\lambda_1}{2}\left(\partial_z u - \varepsilon^2 \partial_x w\right)\left(\tau^{xx} - \tau^{zz}\right)$$

$$= \mu\left(1 + \lambda_2\frac{d}{dt}\right)(\partial_z u + \varepsilon\partial_x w) + \lambda_2\mu\left(\partial_z u - \varepsilon^2 \partial_x w\right)\partial_x u. \tag{5.92}$$

These equations are coupled to the hydrodynamic equations, as before:

$$\partial_x u + \partial_z w = 0, \tag{5.93}$$

$$\varepsilon^2 Re^* \frac{du}{dt} = -\varepsilon^2 \partial_x p + \varepsilon^2 \partial_x \tau^{xx} + \partial_z \tau^{xz}, \tag{5.94}$$

and

$$\varepsilon^2 Re^* \frac{dw}{dt} = -\partial_z p + \partial_x \tau^{xz} + \partial_z \tau^{zz}, \tag{5.95}$$

where $Re^* = Re/\varepsilon^2$ is the *reduced Reynolds number*. Finally, the corresponding scaled boundary conditions at the film surface $z = h(x,t)$ are given by

$$-p_R + \frac{\varepsilon\partial_x h^2 \tau^{xx} - 2\partial_x h \tau^{xz} + \tau^{zz}}{1 + \varepsilon^2(\partial_x h)^2} = \frac{\partial_{xx} h}{(1 + \varepsilon^2)^{3/2}} \tag{5.96}$$

and

$$\tau^{xz}\left(1 - \varepsilon^2(\partial_x h)^2\right) - \varepsilon^2 \partial_x h \bar{\tau} = 0. \tag{5.97}$$

In the last equation, we have introduced the *first normal stress difference* $\bar{\tau}$ which is commonly also denoted by N_1,

$$\bar{\tau} \equiv \tau_{xx} - \tau_{zz}. \tag{5.98}$$

For the expansion, we assume

$$\left(u, w, h, p, \tau^{ij}\right) = \left(u_0, w_0, h_0, p_{R0}, \tau_0^{ij}\right) + \varepsilon^2\left(u_1, w_1, h_1, p_1, \tau_1^{ij}\right) + O\left(\varepsilon^4\right). \tag{5.99}$$

To leading order we have

$$\partial_z u_0 = 0 \tag{5.100}$$

or

$$\tau_0^{xz} = \frac{\lambda_2}{\lambda_1}\partial_z w_0. \tag{5.101}$$

Further, we have

$$\partial_x u_0 + \partial_z w_0 = 0, \tag{5.102}$$

$$\partial_z \tau_0^{xz} = 0 \tag{5.103}$$

$$\partial_z p_{R0} = \partial_x \tau_0^{xz} + \partial_z \tau_0^{zz}, \tag{5.104}$$

and the leading-order boundary conditions at $z = h_0$ are

$$\tau_0^{xz} = 0 \tag{5.105}$$

$$p_{R0} - 2\left(\frac{1}{2}\tau^{xz} - \partial_x h_0 \tau_0^{xz}\right) + \partial_{xx} h_0 = 0, \tag{5.106}$$

$$\partial_t h_0 - w_0 + u_0 \partial_x h_0 = 0. \tag{5.107}$$

Leading-order boundary conditions at the substrate are:

$$w_0 = 0, \qquad \tau_0^{xz} = 0. \tag{5.108}$$

Integrating Eq. (5.103) with respect to z we find the last result to hold generally. With the constitutive equations and the boundary conditions we obtain the plug flow condition $\partial_z u_0 = 0$. Then

$$\left(1 + \lambda_1 \frac{d^*}{dt}\right)\tau_0^{xx} = 2\left(1 + \lambda_2 \frac{d^*}{dt}\right)\partial_x u_0, \tag{5.109}$$

$$\left(1 + \lambda_1 \frac{d^*}{dt}\right)\tau_0^{zz} = 2\left(1 + \lambda_2 \frac{d^*}{dt}\right)\partial_z w_0, \tag{5.110}$$

where $d^*/dt = \partial_t + u_0 \partial_x + w_0 \partial_z$. The pressure at the liquid surface is given by

$$p_{0R} = \tau_0^{zz} - \partial_{xx} h_0. \tag{5.111}$$

The next-order ($O(\varepsilon^2)$) u-momentum equation is

$$Re^* \frac{d^*}{dt} u_0 = -\partial_x p_{0R} + \partial_x \tau_0^{xx} + \partial_z \tau_1^{xz}. \tag{5.112}$$

Using the pressure term and denoting $u_0 = f(x, t)$, we obtain

$$Re^*(\partial_t f + f \partial_x f) = \partial_x \overline{\tau_0} + \partial_x (\partial_{xx} h_0 + \phi') + \partial_z \tau_1^{xz}. \qquad (5.113)$$

Integrating the last equation from 0 to h_0 we find, upon employing the slip-boundary conditions to the next order, $\tau_1^{xz} = f/\beta_s$,

$$h_0 Re^*(\partial_t f + f \partial_x f)$$

$$= \partial_x \int_0^{h_0} dz \overline{\tau_0} - \overline{\tau_0}|_{z=h_0} \partial_x h_0 + h_0 \partial_x (\partial_{xx} h_0 + \phi')$$

$$+ \tau_1^{xz}|_{z=h_0} - \frac{f}{\beta_s}. \qquad (5.114)$$

The next-order tangential stress boundary condition at the film surface is

$$\tau_1^{xz} = \overline{\tau_0} \partial_x h_0. \qquad (5.115)$$

Hence

$$h_0 Re^*(\partial_t f + f \partial_x f) = \partial_x \int_0^{h_0} dz \overline{\tau_0} + h_0 \partial_x (\partial_{xx} h_0 + \phi') - \frac{f}{\beta_s}. \qquad (5.116)$$

From the Jeffreys' equations we obtain an equation for $\overline{\tau_0}$:

$$(1 + \lambda_1 \partial_t + \lambda_1 f \partial_x - \lambda_1 z \partial_x f \partial_z) \overline{\tau_0}$$

$$= 4(\partial_x f + \lambda_2 \partial_t \partial_x f + \lambda_2 f (\partial_x f)^2 - \lambda_2 z \partial_x f \partial_z \partial_x f). \qquad (5.117)$$

We now define a film average S of $\overline{\tau_0}$ as

$$S \equiv \frac{1}{4h_0} \int_0^{h_0} dz \overline{\tau_0} \qquad (5.118)$$

Denoting the rhs of Eq. (5.117) as $G(x, t)$, and integrating with respect to z yields

$$4h_0 S + \lambda_1 \partial_t (4h_0 S) + \lambda_1 f \partial_x (4h_0 S)$$

$$+ \lambda_1 (4h_0 S) \partial_x f - \lambda_1 \overline{\tau_0}|_{z=h_0} (\partial_t h_0 + \partial_x (f h_0)) = h_0 G(x, t) \qquad (5.119)$$

Making use of the kinematic condition to leading order, and of the definition of $G(x, t)$ yields

$$(1 + \lambda_1 \partial_t + \lambda_1 f \partial_x) S = (\partial_x + \lambda_2 \partial_t \partial_x + \lambda_2 f \partial_{xx}) f \qquad (5.120)$$

together with the lubrication equation

$$h_0 Re^*(\partial_t f + f \partial_x f) = \partial_x (4h_0 S) + h_0 \partial_x (\partial_{xx} h_0 + \phi') - \frac{f}{\beta_s}. \qquad (5.121)$$

Several remarks are in order to interpret the result. First of all, the formal result for the lubrication model shows a remarkable similarity to the result for the *Jeffreys model*. Two differences arise: the stress tensor is replaced by a more complex average which involves the film height itself; second, the material law now contains spatial derivatives, which was to be expected, of course. But in Eq. (5.121) only *advective nonlinearities* arise, and not the *corotational nonlinearities*. This implies that the use of upper or lower convective nonlinearities replacing the Jaumann derivative we used, which would lead to the *Oldroyd models* A and B (Byron Bird et al. 1987a, 1987b), will lead to the same result for the two-dimensional thin-film equation as we have obtained it before. Our formal result thus assures us that on the level of complex rheological models with strong slip, the form of the equations remains generally valid.

5.3.2 Confrontation with Experiment: A Phenomenological Modification of the Jeffreys Model

In the previous section we have seen that viscoelastic effects alone cannot be at the origin of the shape change of the dewetting rim profile, while as we know from the previous chapter, a sufficiently strong slip even without viscoelasticity induces a monotonously decaying profile. The question therefore arises in what situation viscoelastic effects can be less ambiguously observed.

So far we have tacitly assumed that the polymer film under study is in a *fluid state*. This is, of course, only the case at elevated temperatures (see Appendix A). For lower temperatures—in particular, ambient temperature—the polymer film is a solid, albeit not a crystalline one, but a *glass*. Therefore, there is a characteristic temperature T_g, below which the film is to be considered glass-like. We will discuss the glassy properties of polymeric films in the subsequent section in some more detail.

Upon approaching T_g, the films certainly become more solid-like, and hence presumably viscoelastic effects will become more prominent. This certainly does not yet say that the model we have chosen, the Jeffreys model, is at all an adequate one. As discussed before, the Jeffreys model has two viscous, and one elastic regime. Are these properties reflected in experiments on polymer films near the glass temperature, and what is observed? These problems have been addressed in a series of papers by G. Reiter, P. Damman, E. Raphaël and their collaborators (Reiter 2001; Reiter et al. 2005, 2009; Gabriele et al. 2006a, 2006b; Damman et al. 2007; Hamieh et al. 2007). We here discuss some of their results, as relevant to our foregoing discussion.

Reiter et al. (2005) performed experiments on polymer films which had been submitted to an 'ageing' process whereby the films were cooled to a temperature of 50 °C for increasing time. Independent of the ageing process, the holes form rapidly (in times <100 s). The number of holes, however, decreases exponentially with ageing time. Dewetting experiments on these samples yield characteristic rim profiles

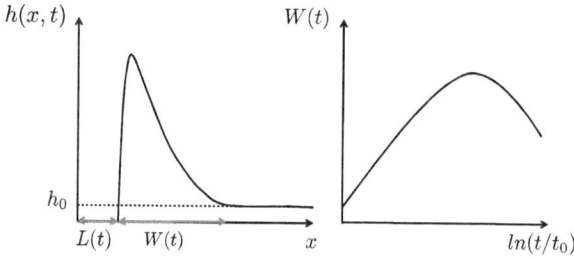

Fig. 5.2 Rim evolution with residual stresses, after Reiter et al. (2005) and Vilmin and Raphaël (2006). *Left*: the monotonous profile of a rim developing in the aged film. Definitions of lengths used in the text, in particular the time-dependent rim width $W(t)$. *Right*: evolution of $W(t)$, schematically after both experiment and theory where an exponent $\alpha = 2/3$ appeared as reasonable choice. The profile goes through a maximum. Note that the time-scale is logarithmic

that decay monotonously towards the flat film. Figure 5.2 shows such profiles taken after different dewetting times. The left graph in the figure explains characteristic lengths of the profiles that can be defined.

The rim width for short times increases logarithmically in time until it reaches a maximum and then decreases again. This behaviour is qualitatively reproducible from the Jeffreys model. Numerical calculations in this case however, have to be based on an important modification: the introduction of a (phenomenological) *non-linear friction* force at the substrate (Vilmin and Raphaël 2005, 2006; Vilmin et al. 2006). Starting point is Eq. (5.65) for the strong-slip viscoelastic thin film model, in which the intertial and the capillary terms have been dropped. By contrast, the linear slippage term has been phenomenologically generalized to a nonlinear function introducing an exponent $0 < \alpha < 1$, which for $\alpha = 0$ covers the strong-slip model. This step yields, in a two-dimensional setting,

$$\zeta \bar{u}^{\alpha} u^{1-\alpha} = \partial_x (hq) \tag{5.122}$$

together with the Jeffreys law and the conservation law of Eq. (5.65). Here, $\zeta = \eta/b$ is the *friction coefficient* of the film on the substrate and $\bar{u}$ a characteristic velocity above which the friction is nonlinear.

This model has been employed for studies of the case of a linear dewetting front. In this case, the radius of an expanding hole is replaced by the *dewetted distance* $L(t)$, whereby $L(t = 0) = 0$. At the film edge, the stress has to fulfill the relation

$$h(L)q(L) = -|S| \tag{5.123}$$

where $S = \sigma_{sv} - (\sigma_{sl} + \sigma_{lv})$ is the *spreading parameter*, see Part I. As initial conditions, one can prescribe a quiescent film of height $h_0 = 0$, $u(w, t = 0) = 0$.

An important difference then remains for the initial condition on the stress q. Either one assumes a relaxed film—as we did in the derivations of the thin film equations—or a *residual stress* with a constant value, q_0.

For short times these equations can be solved analytically, while for long times one has to resort to numerical solutions (Ziebert and Raphaël 2009a, 2009b). Without initial residual stress, we have to solve with $q = 2\eta \partial_x u$

$$\zeta \overline{u}^\alpha u^{1-\alpha} = 2\eta h_0 \partial_x^2 u. \tag{5.124}$$

For $\alpha = 0$, the equation yields an exponential velocity profile

$$u(x, t) = U_0 \exp\left(-\frac{x - L}{\sqrt{2}W_0}\right) \tag{5.125}$$

where L is the dewetted distance, $L(t) = 0$, and W_0 the characteristic rim width. With the boundary condition (5.123) one finds the initial velocity U_0, and has

$$U_0 = \frac{|S|}{(2\eta \zeta h_0)^{1/2}}, \qquad W_0 = \left(\frac{\eta h_0}{\zeta}\right)^{1/2} = (h_0 b)^{1/2} \tag{5.126}$$

with the slip length b. For nonlinear friction, $\alpha \neq 0$, the velocity profile reads as

$$u(x, t) = U_{0,\alpha}\left(1 - \frac{\alpha}{2}\frac{x - L}{\sqrt{2}W_{0,\alpha}}\right)^{2/\alpha} \tag{5.127}$$

for $0 < x - L < 2\sqrt{2}W_{0,\alpha}/\alpha$, and $u(x) = 0$ elsewhere. The corresponding scales of velocity and width are modified accordingly to

$$U_{0,\alpha} = \left[\left(\frac{2-\alpha}{2}\right)\frac{U_0^2}{\overline{u}^\alpha}\right]^{1/(2-\alpha)} \tag{5.128}$$

and

$$W_{0,\alpha} = W_0\left[\left(\frac{2-\alpha}{2}\right)\frac{U_0^\alpha}{\overline{u}^\alpha}\right]^{1/(2-\alpha)}. \tag{5.129}$$

The height profile at short times obeys $\partial_h = -h_0 \partial_x v(x)$ and fulfills for linear and nonlinear friction, respectively, the expressions

$$h(x, t) = h_0\left[1 + \frac{U_0 t}{\sqrt{2}W_0}\exp\left(-\frac{x - L}{\sqrt{2}W_0}\right)\right] \tag{5.130}$$

and

$$h(x, t) = h_0\left[1 + \frac{U_{0,\alpha} t}{\sqrt{2}W_{0,\alpha}}\left(1 - \frac{\alpha}{2}\frac{x - L}{\sqrt{2}W_{0,\alpha}}\right)^{(2-\alpha)/\alpha}\right] \tag{5.131}$$

In both cases the rim builds up with time $h(L(t), t) \propto t$.

If a residual stress is present, the equations need to be solved numerically. Figure 5.2 shows the non-monotonous behaviour of the rim width which develops. As it turns out, both the nonlinear friction term, and a residual stress are needed to reproduce this experimental feature, vindicating the use of viscoelastic models of Jeffreys type, however, with some phenomenological modifications.

5.4 Microscopics of Polymer Films

In this section we will descend—at least at little bit—into the microscopics of thin polymer films. In the previous chapters and sections we have encountered two elements that lead us beyond 'simple' hydrodynamics: the slip length and its variation with surface preparation, and the behaviour of thin films when the glass temperature is approached: here, viscoelastic effects become important. Hence we have to take a look at what happens in the films around $T = T_g$. Slip and glassy behaviour will thus be the topics of the next two subsections.

5.4.1 Microscopics of the Slip Boundary Condition

What happens at the solid substrate on which the polymer film has been coated? How can we influence the slip length? This question has already been addressed some time ago by Brochard and de Gennes (1992). The task at hand is to find expressions that relate the 'macroscopic' slip length to 'microscopic' properties of the substrate. Starting point is the slip formula

$$b = \left.\frac{v}{v'}\right|_0 = \frac{\eta}{\zeta} \tag{5.132}$$

where ζ is a friction coefficient, which we can relate to the shear stress τ via

$$\zeta = \tau/v. \tag{5.133}$$

For an 'ideal', smooth surface, the friction coefficient would be that of a *fluid of monomers*, $\zeta = \zeta_m$. In an entangled polymer melt, the melt viscosity can be huge, such that values of $b \sim 100$ μm are possible; however, such values are rarely encountered. This means that surfaces are usually not ideal; most importantly, it may be that the polymers interact locally with the surface.

One possibility to consider is a surface on which polymers are grafted. Several limits appear possible depending on grafting density and chain length of the grafted polymers; the most interesting regime is the one in which the polymers have long chains that do not overlap, the so-called *mushroom regime*, see Fig. 5.3. At low velocities, the polymers are essentially curled up to spheroids, while at larger flows they become elongated and stretch in the flow. In all situations, however, the grafted chains are not entangled among themselves.

In this case, one can write the imposed shear stress on the polymers as

$$\tau = \eta\frac{dv}{dz} = \zeta_m v + vF, \tag{5.134}$$

whereby the first term describes the weak friction due to monomer wall interactions with

$$\zeta_m \sim \frac{\zeta_1}{a} \sim \frac{\eta_1}{a} \tag{5.135}$$

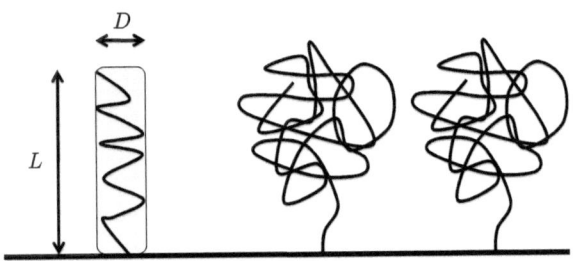

where a is the size of a monomer and η_1 hence the viscosity of a liquid of monomers. The second term is proportional to the elastic force F on the stretched chains of Z monomers per grafted chain, where v is their number per unit area. Writing down this expression, we assume that we treat the chains as non-overlapping, and we can thus already learn something by just considering a single chain with the geometry shown in Fig. 5.3: a chain of diameter D and length L.

The elastic force on such a chain is, for an ideal chain, given by

$$F = \frac{3kT}{R_0^2}L \approx \frac{kT}{D} \tag{5.136}$$

where $R_0 = \sqrt{Z}a$ is the coil size. In a slow flow, for $L < R_0$, the chain is largely curled up to a ball of size R_0 and the friction force associated with it is given by a Stokes law

$$F \approx \eta R_0 v. \tag{5.137}$$

For larger velocities and strong elongation, $L > R_0$, it turns out to be replaced by a similar law, replacing coil size by elongation L:

$$F \approx \eta L v, \tag{5.138}$$

so that one can sum up both regimes by the extrapolation

$$F = \frac{\eta}{\sqrt{1 + (R_0/L)^2}} v. \tag{5.139}$$

Equating now the elastic force with the elongation or friction force, one finds the law

$$v = \frac{v^*}{\sqrt{1 + (R_0/L)^2}} \tag{5.140}$$

where

$$v^* \sim \frac{k_B T}{\eta R_0^2} = \frac{kT}{\eta Z a^2}. \tag{5.141}$$

Thus, for $v \to v^*$, the elongation L formally diverges, which is of course cut off by the finite size of the polymer at a value L^*.

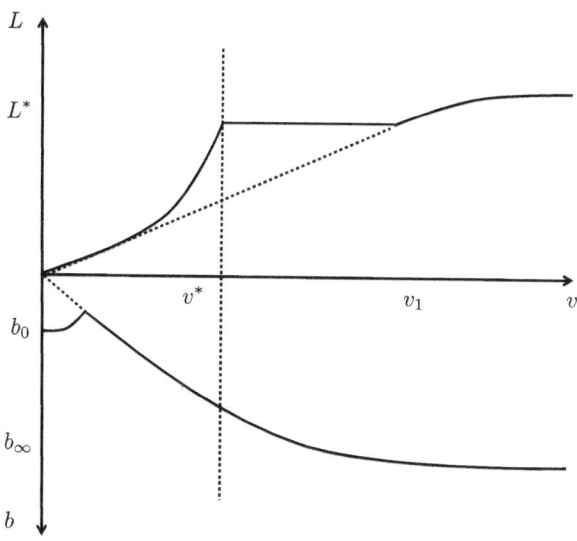

Fig. 5.4 Chain elongation and slip length b as a function of velocity showing the three regimes discussed in the text: entangled, marginal and disentangled regimes. Drawn after Brochard and de Gennes (1992)

But what happens for $v > v^*$? In fact, the chain does not yet disentangle, but enters a so-called *marginal state*. If one assumes a friction law of Rouse type

$$F = Z\zeta_1 v \tag{5.142}$$

which sums up individual contributions of each monomer, a shorter elongation is found by balancing to the elastic force:

$$L_R = L^* \frac{v}{v_1} \tag{5.143}$$

where

$$v_1 = \frac{kT}{Z\zeta_1 R_0^2} L^*. \tag{5.144}$$

The velocity $v_1 \gg v^*$ so that there is a whole interval of speeds in which the chain, after disentangling beyond v^*, shrinks in size L (or broadens to a diameter $D > D^*$) before it completely disentangles into a state with $L > L^*$ when the velocity v_1 is passed. This behaviour is summarized in the top half of Fig. 5.4. The different slip-length regimes predicted by this theory could qualitatively be seen in the experiments by Migler et al. (1993), Léger et al. (2006).

We can now transfer this result to the behaviour of the single chain to the grafted surface, Eq. (5.134), and deduce the slip length behaviour. At very low stresses one finds a friction coefficient

$$\zeta_0 = \zeta_m + v\eta R_0 \approx v\eta R_0. \tag{5.145}$$

The slip length is then given by

$$b_0 = \frac{\eta}{\zeta_0} \approx \frac{1}{\nu R_0}. \tag{5.146}$$

In the marginal state $L = L^*$ the force has a fixed value F^*. Neglecting again the ζ_m-term, possible because of its small size, one has

$$\tau = \tau^* \equiv \nu F^* \approx \frac{\nu k T}{R_0^2} L^*. \tag{5.147}$$

The slip length now is a linear function of velocity

$$b = b(\upsilon) = \frac{\eta}{\sigma^*} \upsilon, \tag{5.148}$$

hence a linear function of velocity υ. In the completely distentangled regime the slip length then is given by the expression

$$b = b_\infty = \frac{\eta}{\zeta_m + \nu \zeta_1 Z} \approx \frac{\eta}{\zeta_m} \tag{5.149}$$

since in this regime ζ_m dominates. The slip length thus reaches a plateau.

To summarize these results, we write $L(\upsilon)$ for the elongation law of the single chain, drawn in the top half of Fig. 5.4. Then we have from the stress law

$$\tau = \zeta_m \upsilon + \frac{\nu k T}{R_0^2} L(\upsilon) = \eta \frac{\upsilon}{b(\upsilon)} \tag{5.150}$$

and solve for $b(\upsilon)$ to obtain

$$b(\upsilon) = \frac{\eta}{\zeta_m + \frac{\nu k T}{R_0^2} \frac{L(\upsilon)}{\upsilon}}. \tag{5.151}$$

The result is shown in the lower half of Fig. 5.4. The slip length has a non-monotonous behavior at low velocities before it enters first a linear regime and then saturates.

The results from this section certainly can convince us that the description of thin polymer films necessitates a proper (microscopic) modeling of the substrate-film interaction, and for the case we encountered in our viscous flows this has yet to be done. It may be expected that very recent experimental insights will further help in this respect, see, e.g. Gutfreund et al. (2011).

In the next section, we go on to discuss 'bulk' properties of thin films: a bulk which is strongly affected also by the free surface.

Fig. 5.5 Sketch of the behaviour of the glass transition temperature T_g as a function of film thickness h, as experimentally observed for polystyrene (PS) films with various experimental techniques. The data show a power-law decrease, see Eq. (5.152) (after Roth and Dutcher 2005)

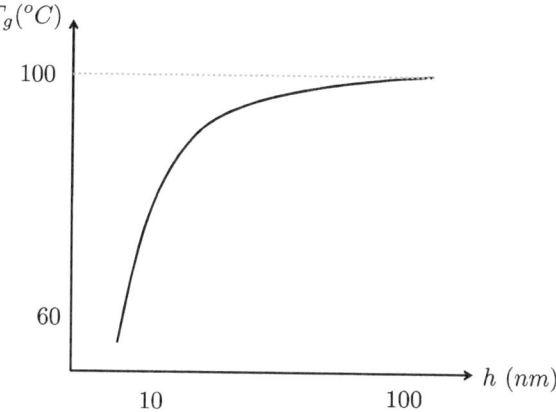

5.4.2 *Polymers are Glasses: T_g of Thin Films*

In this final subsection we enter contested ground—but in fact, we do have a hard time in really touching it. As we had seen before, polymer films are usually not crystalline in their solid state, and hence the transition between a solid and a fluid state is glassy territory. There are numerous experimental observations relating to the glass temperature of thin polymer films, and many theoretical ideas have been put forward to explain them. 'Idea' is written here with a purpose, since at present there is no consistent theory of the observed phenomena. This subsection can thus neither fulfill the task to review all this material (which would easily fill a whole volume on its own), and even less so can it present the 'ultimate theory' of T_g in thin polymer films. In the absence of a decent theory, all we will do is to review experimental material and get an idea of the physics that might play a role.

Figure 5.5 sketches the general trend that is experimentally observed for the *glass temperature* of thin polymeric films in the thickness range of 100 nm and less. These data have e.g. been assembled by Roth and Dutcher (2005) who united results obtained with a variety of different methods, and for films of different molar masses M_w. Although the experiments still show a substantial scatter in T_g below thicknesses of 100 nm, the general trend of a reduction of T_g is followed.

In ellipsometry experiments the glass transition temperature is obtained from monitoring the film thickness. Upon heating, the films expand thermally, and this expansion usually encounters a discontinuity, visible in a soft kink in the expansion curve. Since the phenomenon is not sharp, the precise measurement of T_g has by itself a substantial intrinsic error. Typically, the data can be fitted to an empirical expression

$$T_g(h) = T_g^{bulk}\left[1 - \left(\frac{h_0}{h}\right)^{\delta}\right]$$
(5.152)

where values for δ lie typically in the range $1 \leq \delta \leq 2$.

We are here only concerned with *supported films*, but of course one can study the same problem also for *free-standing films*; indeed, T_g reductions turn out to be much larger in this case. The presence of the substrate thus does influence T_g, but for fluids that only weakly interact with the substrate, such as polystyrene (PS) the effect appears not to be essential. This is not true for other systems (i.e., combinations of films and substrates), which can even lead to a reversal of the behaviour, i.e. an increase rather than a decrease in T_g.

The reduction of T_g must be related to some mechanism (or several) related to chain motion. Early models have been built on this premise (de Gennes 2000; Herminghaus 2002; Herminghaus et al. 2003). There is an ongoing debate whether the effect affects the whole film uniformly, or whether a more mobile layer is present. The situation is clearly complex also due to the different ways to probe the glassy behaviour. Another quantity of relevance is the viscosity of the thin film. We had seen in the previous chapter that the dynamics of thin film rupture critically depends on viscosity, as it sets the timescale, and we had invoked fluctuation effects to cure that discrepancy. There is, however, also a dependence of viscosity on film thickness. This is e.g. reflected in the viscosity measured in dewetting experiments by the analysis of power spectral densities (Yang et al. 2010) or by ellipsometry measurements in cooling experiment on supported PS-films (Fakhraai and Forrest 2008).

Yang et al. (2010) have recently put forward the following reasoning in favor of the existence of a mobile surface layer. Starting point is the temperature-dependence of the viscosity of a glass-forming liquid is related to its glass temperature by the *Vogel-Fulcher-Tammann*, VFT, relation

$$\eta(T) = \eta_\infty e^{\frac{B}{T - T_K}} \tag{5.153}$$

where B is a constant and T_K is the *Kauzmann temperature*. The latter is defined as the temperature at which the *configurational entropy* of an ergodic, supercooled liquid vanishes. For the PS used in the experiments by Yang et al. (2010), $T_K = 288$ K, while $T_{g,bulk} = 337$ K; $B = 1620$ K. The experimental data on $T_g(h)$ for the experiment can then be fitted by the curve

$$T_K(h) = \frac{T_{g,\infty}}{(1 + h_0/h)} - \Delta T \tag{5.154}$$

where $h_0 = 0.6 \pm 0.1$ nm, and $\Delta T = 48 \pm 3$ K, as in bulk. Therefore, also from the viscosity measurements, the T_g reduction can be found as from thermal expansion measurements.

If the experimental data of Yang et al. (2010) are plotted as η/h^3 vs $1/T$, the *Arrhenius curve*

$$\frac{3\eta}{h^3} = A e^{\frac{E}{RT}} \tag{5.155}$$

with the parameters $A = 165 \pm 7$ Pa s m^{-3}, and $E = 185 \pm 3$ kJ/mol, all data for the thinner films within the range of thicknesses 2.3 nm $\leq h \leq 9$ nm can be assembled on a single scaling curve, while the bulk data deviate from this line. If η is plotted vs. $1/T$, the bulk data assemble on the bulk curve, while the all other data lie on straight lines which follow the relation of an *effective viscosity*

$$\eta_{eff} \approx \frac{h^3}{3(M_t + M_b)} \tag{5.156}$$

where the denominator contains two mobilities of a bulk and a top-layer. This relation follows from the assumption of a bilayer model for the film with a bulk-like inner layer and a more mobile surface layer. The experimental results are then explained by the fact that for the thin films the mobility of the thin surface layer dominates, hence

$$\eta_{eff} \approx \frac{h^3}{3M_t}. \tag{5.157}$$

In addition, the model by Yang et al. (2010) requires a finite thickness of the mobile surface layer which does not require a critical thickening as proposed in alternative models.

This ends our discussion of microscopic effects. Those interested in still better founded theoretical models, either from the point of view of hydrodynamic theories, or from more microscopic ('kinetic') approaches, are directed to Appendix D.

Chapter 6
Conclusions and Outlook

In this book we have developed the body of the theory of wetting and dewetting, and in particular discussed the fundamental aspects of dewetting thin polymer films, based on methods from statistical physics and hydrodynamics. Although the presentation has by choice been restricted to fairly clear-cut situations—see the motivation in the Introduction—we have seen that even then one is rapidly pushed to resort to additional, often phenomenological approaches.

Even within the chosen limitations, we have not discussed everything that would have been possible. The list of topics we have not touched is indeed long. In particular, there are numerous situations in dewetting that were discussed previously by F. Brochard, P.G. de Gennes and their collaborators. To just name a few:

- dewetting under gravity control: (Brochard et al. 1988);
- liquid-liquid dewetting (Brochard-Wyart et al. 1993);
- dewetting between a porous solid and a rubber (de Gennes 1994; Brochard-Wyart and de Gennes 1994b; Martin and Brochard-Wyart 1998);
- dewetting of a stratified liquid (Ausserré et al. 1995);
- inertial dewetting (Brochard-Wyart and Buguin 1999);
- Cerenkov dewetting (Martin et al. 2002)

... a long list, and *still* incomplete. Maybe the (somewhat) more rigorous approach pursued in this book will lure some readers into a more detailed investigation of the above topics.

But in this last chapter, our look goes ahead into *other* directions, for which even less is known and established in terms of the mathematical description. In the following outlook section I have chosen to briefly discuss five of these topics. They are: finite geometries, evaporation, metallic films, polymers under external fields and the dynamics of the cytoskeleton. In going from the first to the last example we will pass from equilibrium to non-equilibrium phenomena.

R. Blossey, *Thin Liquid Films*, Theoretical and Mathematical Physics,
DOI 10.1007/978-94-007-4455-4_6, © Springer Science+Business Media Dordrecht 2012

Fig. 6.1 *Top*: Droplets attached to the corrugations of a plastic cup. *Bottom*: An increasing complexes is expected in going from a virtually infinite wedge (**a**) to a periodic pattern of wedge-shaped grooves (**c**). The shape (**b**) is a compromise between both. Reprinted with permission from Brinkmann and Blossey (2004). Copyright by the European Physical Society

6.1 Finite and Structured Geometries

In all situations covered so far in the book, except for the liquid dewetting phenomenon discussed in the Introduction, the boundary conditions have been simple: we could essentially ignore them, since we looked at infinite films. In every experiment, of course, boundary conditions matter, and indeed this can be turned around, as they can also be used to trigger the phenomenon in the first place. This becomes most important when the systems under study acquire more complex geometries than a mere flat substrate.

Two main directions have been developed in recent years without having exhausted the subject as the variations are clearly manifold. The first is to look at the instability of thin films (again mostly polymer) on structured substrates. The most common examples are substrates containing stripes of different wettability. The dewetting phenomenon then has to accommodate a new length-scale, set by the wettability contrast (Kargupta et al. 2000; Kargupta and Sharma 2002; Ledesma-Aguilar et al. 2009).

A second line consists of geometry-dominated effects, in which liquids are placed on topographic substrates. We illustrate this line of research with the following example, taken from the real world of coffee breaks. Figure 6.1 shows an ordinary plastic cup which bears a corrugation. Some typical topographies are indicated at the bottom of the figure (Brinkmann and Blossey 2004). The step geometry (c) in Fig. 6.1 has been analyzed mathematically with analytical and numerical means. In this case (as in many others) interfacial interactions are ignored: the study is thus applicable to droplets in the micron range where both the gravitational and dispersion interactions are negligible. Figure 6.2 shows the studied geometry of a drop attached to a step with opening angle α.

Fig. 6.2 *Top*: Liquid cross sections in a virtually infinite wedge at different contact angles, going from a spreading to a droplet configuration from (**a**) to (**c**). *Below*: step edge wetted by a drop. Several angles are indicated to describe the liquid geometry. *Bottom*: two liquid configurations at the step: a 'liquid cigar' (*top*) and a 'blob'. The liquid does not spill over to the upper surface. Reprinted with permission from Brinkmann and Blossey (2004). Copyright by the European Physical Society

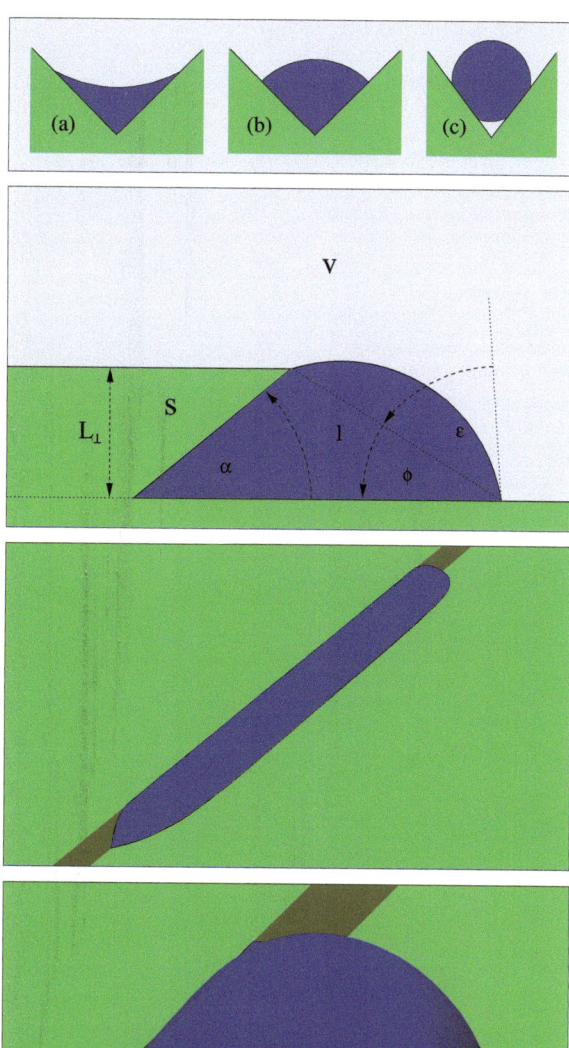

The computational task is the minimization of the interfacial free energy

$$\mathcal{F} = \sigma_{lv} A_{lv} + (\sigma_{ls} - \sigma_{vs}) A_{ls} \tag{6.1}$$

under conditions of mechanical equilibrium, i.e. the fulfillment of the *Laplace equation*

$$2\sigma_{lv}\kappa = p_l - p_v \tag{6.2}$$

Fig. 6.3 *Top*: Morphology diagram at a fixed step angle $\alpha = \pi/2$ as a function of contact angle and volume. High volumes and contact angles favor the formation of blobs. At the *solid line*, both configurations have the same interfacial free energy. *Bottom*: Morphology diagram for high volumes, $V/L_\perp^3 \to \infty$. Cigar-shaped droplets are only locally stable configurations. Reprinted with permission from Brinkmann and Blossey (2004). Copyright by the European Physical Society

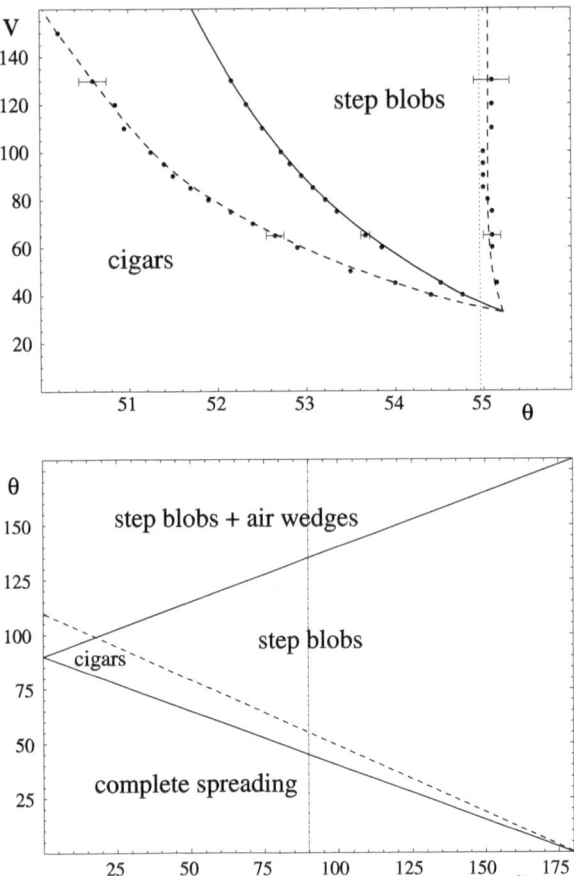

where κ is mean curvature, see Chap. 4. Further, the Young-Dupré equation has to be fulfilled at the contact line. Figure 6.2 displays the two characteristic emerging topologies: a 'blob', a droplet localized at the step, and a 'cigar', a droplet which, while also attached to the step, acquires an elongated shape along the step. Figure 6.3 summarizes the calculated profiles in terms of *morphology diagrams*.

6.2 Evaporation

Evaporation effects in wetting and dewetting phenomena have gained prominence due to the famous 'coffee spot effect' (Deegan et al. 1997; see also Parisse and Allain 1996), see Fig. 6.4. An evaporating droplet containing dispersed particles leaves a fairly regular ring on a surface—indeed an everyday phenomenon, so much that it takes a real experimentalist to notice it in the first place. The explanation of the effect lies in that evaporation is fastest at the contact line which leads to a net flux of

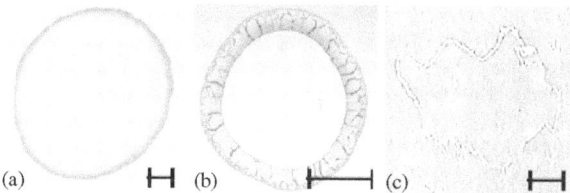

Fig. 6.4 (**a**) Coffee stain; (**b**) dried colloidal microsphere; (**c**) salt deposit. The scale bar corresponds to approximately 1 cm. Reprinted with permission from Deegan et al. (2000). Copyright by the American Physical Society

particles towards the contact line where they become deposited. Mathematically, this can be shown by invoking an analogy to electrostatics. An electric field at the contact line diverges, and due to the mathematical identity of the Poisson and diffusion equation this leads to a divergent particle flux at this point.

If the liquid making up a film is volatile, evaporation can also trigger film rupture. In some cases evaporation effects can be modeled by a temperature gradient across the film which adds a contribution to the effective interface potential; a detailed theoretical discussion of this case are given in Burelbach et al. (1988); Oron et al. (1997). From a conceptual point of view it becomes important to include exchange with the vapour phase, a point we could neglect in all of our discussions on polymer films.[1] Quite generally, the contribution due to evaporation in the dynamics of droplets or films are highly nonlinear and, in particular, *very hard* to control experimentally. A quantitative comparison between experiment and theory is therefore a real challenge in such situations.

Evaporation effects in thin films have been discussed in the context of *self-assembly* (Ohara et al. 1997; Ohara and Gelbart 1997; Archer et al. 2010; Samid-Merzel et al. 1998; Thiele et al. 1998). One example for this phenomenon is the assembly of *porphyrin rings* or *wheels* (Schenning et al. 1996). Porphyrins are biologically relevant molecules that form complex ring-shaped structures, hence an understanding of their self-assembly is of importance. In evaporating solutions they have also led to ring-shaped objects—raising the question whether the driving force for the formation of these structures resides in the chemistry of the molecules or the physics of the thin films. Their assembly turned out to be rather a mixed dewetting-evaporation effect than a molecular one. The ring domains left on the substrate after evaporation of the film could be monitored by NSOM (Fig. 6.5). A mathematical model for the evaporating thin film could be developed which includes a highly nonlinear disjoining pressure due to the 'recoil' caused by particle loss of the film—this is the dominant effect in this system. Based on experiments under different conditions, a semi-quantitative morphology diagram could be deduced based on a variation of porphyrin concentration and evaporation times. Except for the basic effect, the link between experiment and theory is still poor, however: it would still be real

---

[1]In fact, this is also not entirely correct since in polymer films, at least in their initial stages when volatile solvents are also present. We cautiously avoided to discuss this aspect.

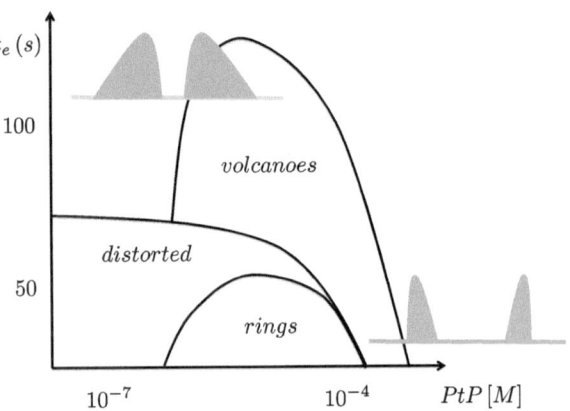

Fig. 6.5 Morphology diagram obtained from NSOM imaging on a 10^{-6} M PtP/CHCl3 sample on glass prepared under ambient conditions, drawn after (Latterini et al. 1999). Indicated are ring-shaped structures and less developed 'volcanoes' whose appearance depends on concentration of PtP molecules

challenge to quantitatively model and prove in experiment the different structure formation processes and assemblies.

6.3 Metallic Films

Metallic films were indeed first used by Bischof et al. (1996), Herminghaus et al. (1998) in order to study the phenomenon of spinodal dewetting, which was controversial at the time in polymeric films, and another system was therefore nice to have. From an experimental point of view, metallic films were of interest as a complementary system to polymers, since the involved surface energies and the viscosity are significantly different between both systems.

Bischof et al. performed experiments on gold, copper and nickel films that were thermally evaporated on fused silica substrates, at layer thicknesses of 25 to 50 nm, with a chromium layer of a few monolayer thicknesses interlaced. The films were then irradiated with laser light to induce dewetting. The experimental observation was the co-occurrence of two hole types which were attributed to resulting from heterogeneous nucleation and spinodal dewetting. Since no theoretical modeling effort was undertaken, these first results were certainly not more than qualitative. Herminghaus et al. (1998) then provided a quantitative analysis in comparison to polymer and liquid crystal films, based on the Minkowski functionals.

Meanwhile these initial results have developed into techniques to tailor nanostructures by making use of surface-dominated phenomena. The technique used has been called *Pulsed-laser-induced dewetting, (PLID)* (Fowlkes et al. 2010). Beyond being able to produce nanostructured surfaces, they also allow to investigate fundamental phenomena like liquid instabilities of Rayleigh-type which we discussed before (Kondic et al. 2009). Figure 6.6 shows two pseudo one-dimensional Nickel (Ni) wires of different thickness exposed to laser irradiation. The bottom figure shows how these wires disintegrate into droplets, and in particular details of the detachment process can be captured. Metallic films or wires are thus both of application

Fig. 6.6 Rupture of irradiated Ni nanowires. The detail shows the detachment of a drop. Reprinted with permission from Kondic et al. (2009). Copyright by the American Physical Society

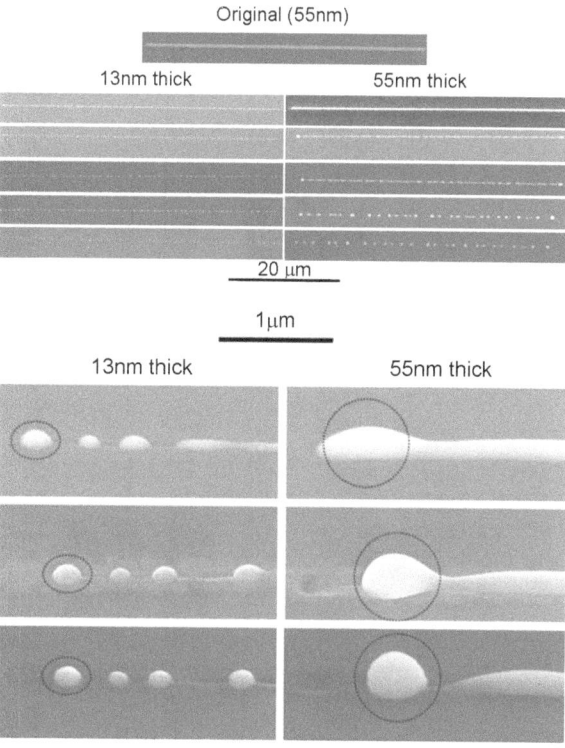

interest and represent yet another model system for which theoretical developments are possible, as witnessed by a number of recent papers that have appeared on the topic (Atena and Khenner 2009; Kondic et al. 2009; Trice et al. 2007, 2008; Wu et al. 2010). For the mathematical modeling, e.g., thermal gradients due to laser irradiation have to be taken into account, which goes beyond the mere interfacial description based on capillary and dispersion interactions.

6.4   Polymers Under External Fields

Polymer films can display intriguing instabilities when they are placed under thermal gradients or electrostatic fields; these processes are of particular interest due to their technological relevance in the production of nanostructured surfaces. In the case of polymers, the application of an electric field can still be covered, to lowest order, by an interface description in which, alongside with the dispersion contribution, an *electrostatic pressure* (Lin et al. 2002; Schäffer et al. 2001)

$$p_{el} = -\varepsilon_0 \varepsilon_p (\varepsilon_p - 1) E_p^2 \tag{6.3}$$

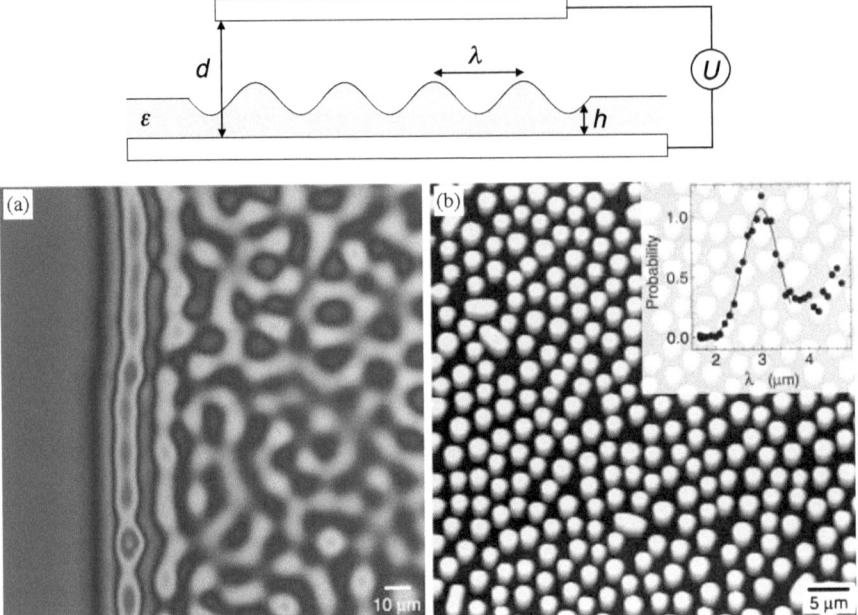

Fig. 6.7 Electrostatic instability of polymer films. *Top*: Schematic representation of a capacitor device used to induce the electrostatic instability. *Bottom*: instability arising in the polymer film. (**a**) Optical microscopy, (**b**) AFM. Reprinted with permission from Schäffer et al. (2001). Copyright by the European Physical Society

is added, where the index p alludes to the polymer. The electric field in the polymer film is given by

$$E_p = \frac{U}{\varepsilon_p d - (\varepsilon_p - 1)h} \tag{6.4}$$

where U is the applied voltage and d the gap between the two electrodes, see Fig. 6.7 (top). The electrostatic pressure acting at the polymer-air interface causes an instability in the film. This instability has a well-defined wavelength, which from the model can be determined as

$$\lambda = 2\pi \left(\frac{\sigma U}{\varepsilon_0 \varepsilon_p (\varepsilon_p - 1)^2} \right)^{1/2} E_p^{-3/2}, \tag{6.5}$$

and is clearly visible from experiment, see Fig. 6.7 (bottom).

This instability can be put to use by employing a structured electrode. The polymer film will then start to replicate the pattern imposed on the surface (Schäffer et al. 2000). Similar patternings can be achieved by the application of a thermal gradient across the film. The origin of the underlying instability mechanism destabilizing the film is still subject to debate (Dietzel and Troian 2009; Schäffer et al. 2003).

6.5 Active Polar Gels

Polymer films have, as we have seen, viscoelastic properties. There are related systems which show viscoelasticity, but they are vastly more complex: cells crawling at surfaces. The mathematical methods for thin films and polymers have been used to develop first models for such systems. The models run under the name of *active polar gel*: active, because the motion of a cell on a substrate is not like a dewetting film which moves and ruptures under the action of 'passive' forces; it is an energy-consuming polymer dynamics of the cytoskeleton that causes it. 'Polar' because the polymers are oriented (actin) filaments to produce coordinated motion; and 'gel', since it is a viscoelastic medium.

Kruse et al. (2005) have developed a hydrodynamic theory of active polar gels. The starting point is the consideration of the stresses in the polar gel, which bears in fact some similarities to liquid crystals. The proper generalization is the expression (Joanny and Prost 2009)

$$\tau = 2\eta u + vh - \zeta \Delta \mu. \tag{6.6}$$

In this expression, the first term is the usual liquid term and the specific terms for the polar gel are the two subsequent terms. In the second, h is an orientational field which is related to the polarization field of the actin molecules via the differential equation

$$h = K \Delta P \tag{6.7}$$

where K is the *Franck constant*. The polarization field p is thus the source of the field h, and it has a time-dependence (a flux) given by

$$\frac{\partial P}{\partial t} = \frac{h}{\epsilon} + v' u \tag{6.8}$$

where ϵ is a rotational viscosity. The two coefficients v and v' are related to each other by Onsager symmetry relations.

Finally, the last term expresses the active property of the polar gel as it is determined by ATP-hydrolysis

$$\Delta \mu = \mu_{ATP} - \mu_{ADP} - \mu_{P_i} \tag{6.9}$$

and ζ is an activity coefficient, a material property of the cytoskeleton characterizing the properties of the molecular motors.

Based on these elements, a full hydrodynamic theory of active polar gels has been developed (Kruse et al. 2005) which bears strong similarities to liquid crystal theory, as alluded to. Another key point in the development of the hydrodynamic theory is the presence of corotational derivatives in the treatment of polarizations. For the full body of equations, we refer to Kruse et al. (2005), and to the introductory paper by Joanny and Prost (2009).

6.6 Further Reading

This brief outlook ends this book. We close it with some hints at some general literature for further study. A classic on wetting as it stood in the mid-eighties is the review article by de Gennes (1985). The results of P.G. de Gennes and his collaborators are covered in the richly illustrated introductory book (de Gennes et al. 2005). Two recent reviews cover both the wetting aspects (Bonn et al. 2009) and the dynamics of thin films (Craster and Matar 2009).

Appendix A
Polymeric Thin Films

In Appendix A we collect basic information on the physical chemistry of the polymers that are used to make thin films, on the substrates and on the experimental methods.

Polystryrene   The most commonly used polymer is *polystyrene*, PS, $[C_8H_8]_n$, see Fig. A.1. PS is one of the most widely used plastic materials. An important aspect for the polymer properties is the structural arrangement of the monomers. PS has different *isomers* which vary in their *tacticity*, the placement of the sidegroups. In isotactic polystyrene, all sidegroups lie on the same side, in syndiotactic polystyrene they change position periodically, while in atactic polystyrene, their positioning is random. Most experiments described in this book were done with atactic polystyrene.

Tacticity influences, e.g., the glass temperature T_g of the polymer.

PMMA/PDMS   Other polymers used in thin-film experiments are PMMA, Poly(methyl-methacrylate) $[C_5O_2H_8]$ and   PDMS,   polydimethylsiloxane, $[C_2H_6OSi]_n$ (see Fig. A.2).

The molecular weight M_n of a single polymer chain is determined by the polymerisation degree n and the molar mass of a monomer via

$$M_N = N M_{mono}. \tag{A.1}$$

For polystyrol, $M_{mono} = 104$ g/mol. If we pass from a single molecule to a distribution of chains, one can introduce the average molar mass

$$M_n = \sum_N n_N M_N \tag{A.2}$$

where n_N is the number of molecules of polymerisation degree N. Further, one defines

$$M_N = \sum_N w_N M_N \tag{A.3}$$

where $w_N = n_N N_N / M_n$ is the mass proportion of molecules of degree N. The *polydispersity index* is then given by M_w / M_n.

R. Blossey, *Thin Liquid Films*, Theoretical and Mathematical Physics,
DOI 10.1007/978-94-007-4455-4, © Springer Science+Business Media Dordrecht 2012

Fig. A.1 Chemical formula
of PS

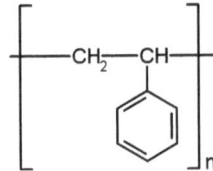

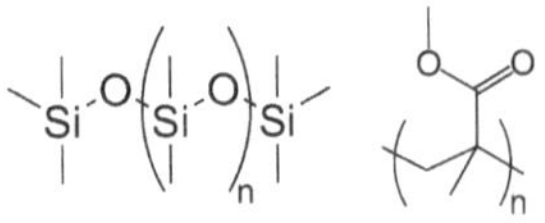

Fig. A.2 PDMS and PMMA

A.1 The Substrates and the Surface Coatings

As substrates, most often Si-wafers are used, usually coated with oxide layers of variable thickness. As discussed in the book, the thickness of the oxide layer is a means to influence the effective interface potential.

Of prime importance are the surface coatings on top of these wafers, since they affect the hydrodynamic boundary condition at the wall. Typically, the coatings are made with *self-assembled monolayers* based on *silanes* (Brzoska et al. 1994). These have an endgroup that covalently binds to the silicon oxide. If a sufficient chain length of the silane is available, brush-like surfaces can be generated. For this purpose, typically n-alkyltrichlosilanes with $n > 10$ are used. The most typical example is octodecyltrichlorosilane $CH_3(CH_2)_{17}SiCl_3$, OTS, which bears a chain composed of 18 carbon atoms. Dodecyltricholorsilane, DTS, bears 12 carbon atoms in the chain. PDMS can also be used to coat the substrate, in conjunction with an elastomer. The surface density R of a coating is related to the average number of grafted chains per unit surface, m, through $m = R/a^2$, with a the size of a monomer, $a = 0.5$ nm for PDMS.

A.2 Preparation and Measurement of Thin Films

Polymer films are usually produced by *spin coating*. In this procedure, a droplet of the polymer solution (polymer and solvent, the latter of which evaporates) is put on a rotating surface. This procedure works if the solvent wets the substrate; if it does not, the film needs to be spin coated on a wettable substrate and brought onto the unfavorable substrate e.g. by *floating*. In this process, the coated solid film is brought onto a water interface and transferred to the wafer.

Measurement of thin films are done with *ellipsometry* and, most importantly, with *atomic force microscopy*, AFM. The first technique relies on refraction of light at the interfaces, while the second brings a atomic-scale probe into interaction with the substrate.

Appendix B
Minkowski Measures

In the analysis of the dewetting images we have made use of the *Minkowski measures* from the field of *integral geometry* (Mecke and Stoyan 2000). Their use is based on *Hadwiger's theorem* which classifies the isotropic measures on compact convex sets in d-dimensional Euclidean space. Every measure can be expressed as a linear combination of $d + 1$ fundamental measures. In two dimensions, there are three possible measures of this type, one corresponding to area, a second corresponding to perimeter, and a third corresponding to the *Euler characteristic*.

In our application, we have to consider two-dimensional height surface in three dimensions, the film height $h(x, t)$. Of course, one could analyse the patterns arising in a rupturing thin film also by Fourier analysis, but this would only give information on the second-order moments. It is therefore advantageous to define contour lines, just as in a geographic map

$$\mathcal{C}_h = \left\{ x \in \mathbb{R}^2 \mid h(x, t) = h \right\}. \tag{B.1}$$

$\mathcal{C}_h$ consists of closed planar lines, or loops, which can be parametrised by a continuous variable s. The Minkowski measures, or more precisely in our context, *Minkowski functionals* are defined as integrals of the profile $h(x, t)$. As stated above, the first of them is the area F (note that we drop the time-dependence in the following)

$$F(h) \equiv \int_A d^2x \, \Theta\big(h(x) - h\big) \tag{B.2}$$

where

$$\frac{\partial F}{\partial h} = -\int_A d^2x \, \frac{1}{|\nabla h(x)|} \delta\big(|x - \mathcal{C}_h|\big) = -\int_A d^2x \, \delta\big(h(x) - h\big). \tag{B.3}$$

The second is the boundary length

$$U(h) = \int_A d^2x \, \delta\big(|x - \mathcal{C}_h|\big) = \int_A d^2x \, |\nabla h(x)| \delta\big(h(x) - h\big), \tag{B.4}$$

R. Blossey, *Thin Liquid Films*, Theoretical and Mathematical Physics,
DOI 10.1007/978-94-007-4455-4, © Springer Science+Business Media Dordrecht 2012

and finally the Euler characteristic

$$\chi(h) = \int_A d^2x\kappa(x)\delta\big(|x - C_h|\big) = \int_A d^2x\kappa(x)|\nabla h(x)|\delta\big(h(x) - h\big) \qquad \text{(B.5)}$$

where $\kappa(x)$ is the curvature of the contour line at point x.

In our application, we have normalized the functionals in a different manner. This is guided by an application to *Gaussian random fields*. For these, one has

$$F(h) = F_0\left(1 - \frac{1}{\sqrt{2\pi}\sigma}\int_{-\infty}^{h} dx e^{-\frac{(s-h_0)^2}{2\sigma^2}}\right), \qquad \text{(B.6)}$$

$$U(h) = U_0 e^{-\frac{(s-h_0)^2}{2\sigma^2}}, \qquad \text{(B.7)}$$

$$\chi(h) = \chi_0 h e^{-\frac{(s-h_0)^2}{2\sigma^2}}. \qquad \text{(B.8)}$$

If one now normalizes according to

$$s(h) = -\frac{\partial F}{\partial h}\frac{1}{U(h)}, \qquad \text{(B.9)}$$

$$u(h) = \ln U(h), \qquad \text{(B.10)}$$

and

$$\kappa(h) = \frac{\chi(h)}{U(h)} \qquad \text{(B.11)}$$

one ends up with the simple algebraic expressions

$$s(h) = s_0 \qquad \text{(B.12)}$$

$$u(h) = u_0 - h_1(h - h_0)^2 \qquad \text{(B.13)}$$

and

$$\kappa(h) = \kappa_1(h - h_0)^2. \qquad \text{(B.14)}$$

Appendix C
Numerics of the Thin-Film Equation

In this appendix we will give, for completeness, a schematic description of the algorithm developed to solve the thin-film equation numerically which was used for the results in Part I. We summarize the essentials of the work by Grün and Rumpf (2000) and Becker and Grün (2005), and refer for a full description to these papers.

The Problem   We consider the discrete solution to the problem

$$h_t - \nabla\big(m(h)\nabla p\big) = 0, \quad \Omega \times (0, T), \tag{C.1}$$

where

$$p = -\Delta h + \Phi'(h). \tag{C.2}$$

The boundary conditions are given by the *no-flux conditions*

$$\partial_\nu h = \partial_\nu p = 0, \quad \text{on } \partial\Omega \times (0, T) \tag{C.3}$$

and the initial data

$$h(\cdot, 0) = h_0(\cdot), \quad \text{on } \Omega. \tag{C.4}$$

The mobility is given by the function

$$m(s) = s^n \tag{C.5}$$

whereby, in general, $n \in (0, \infty)$ is allowed. Further, the interface potential $\Phi(h)$ is assumed to be bounded from below and supposed to be a composed as

$$\Phi(h) = \Phi_+(h) + \Phi_-(h) \tag{C.6}$$

where Φ_+ is a convex, Φ_- a concave function, C^1 over $\mathbb{R}$.

The Triangulation   The domain $\Omega \in \mathbb{R}^N$ for $N = 1, 2$ needs to be triangulated with simplicial elements, which for $N = 2$ are supposed to be rectangular. One then defines by V^l the space of continuous linear functions on each element E of a triangulation $\mathcal{T}^l$. A function $V \in V^l$ is then uniquely defined by its values on the

R. Blossey, *Thin Liquid Films*, Theoretical and Mathematical Physics,
DOI 10.1007/978-94-007-4455-4, © Springer Science+Business Media Dordrecht 2012

nodes of the triangulation $\mathcal{T}^l$. A set of basis functions dual to the set of nodal points is given by $\phi_j \in V^l$ with $\phi_j(x_i) = \delta_{ij}$. Finally one defines a scalar product

$$(\Theta, \Psi)_l \equiv \int_\Omega \mathcal{I}_l(\Theta\Psi) \tag{C.7}$$

with $\mathcal{I}_l : C^0(\Omega) \to V^l$ as an interpolation operator

$$\mathcal{I}_l = \sum_{j=1}^{\dim V^l} u(x_j)\phi_j. \tag{C.8}$$

Time Discretization   The time interval $I \equiv [0, T]$ is subdivided in intervals $(t_k, t_{k+1}]$ with $t_{k+1} = t_k + \tau_k$ for increments $\tau_k > 0$ and $k = 0, 1, \ldots, K - 1$. An implicit backward Euler scheme for Eqs. (C.1), (C.2) is given as follows. Writing the initial condition in triangulated form $H^0 = \mathcal{I}_l h_0 \in V^l$, functions $H^{k+1} \in V^l$ and $P^{k+1} \in V^l$, for all k, need to be determined which fulfill the discrete equations

$$\left(H^{k+1} - H^k, \Theta\right)_l + \tau_k\left(M_\sigma\left(H^{k+1}\right)\nabla P^{k+1}, \nabla\Theta\right) = 0, \tag{C.9}$$

$$\left(\nabla H^{k+1}, \nabla\Psi\right) + \left(\Phi'_+\left(H^{k+1}\right), \Psi\right)_l + \left(\Phi'_-\left(H^k\right), \Psi\right)_l = \left(P^{k+1}, \Psi\right)_l \tag{C.10}$$

for all $\Theta, \Psi \in V^l$. The destabilizing term Φ'_- is discretized explicitly in time, whereas the stabilizing term Φ'_+ is discretized implicitly; this is needed in order to bound the energy at time T by the energy of the initial data. We now close by sketching the construction of the discrete mobility, M_σ, and of ∇H.

Constructing the Mobility   The discrete mobility M_σ is a continuous mapping

$$M_\sigma : V^l \to \bigotimes_{k=1}^{|\mathcal{T}_l|} \mathbb{R}^{N \times N} \tag{C.11}$$

from a function $H \in V^l$ to an associated field of $(N \times N)$-matrices (symmetric and positive definite), each of which is constant on a triangulation element. Further,

$$\nabla H = M_\sigma(H)\nabla\mathcal{I}_l G'_\sigma(H), \tag{C.12}$$

where

$$G_\sigma(s) \equiv \int_\Lambda^s dr\, g_\sigma(r) \tag{C.13}$$

with

$$g_\sigma(s) = \int_\Lambda^s dr\, [m_\sigma(r)]^{-1}. \tag{C.14}$$

G_σ is by construction non-negative and convex, and m_σ is an approximation to the continuous mobility m, e.g., $m_\sigma \equiv m(\max(\sigma, h))$ with $\sigma \ll 1$.

The matrix M_σ is calculated, for $N = 2$, by mapping it with the help of an affine mapping A to a reference triangle $\widehat{E}$ with nodes at $(0, \alpha e_1, \alpha e_2)$. The 2×2-matrix is then given by

$$\widehat{M} = \left(\delta_{ij}\varrho\left(H(0), H(\alpha_i e_i)\right)\right)_{i,j=1,2} \tag{C.15}$$

with

$$\varrho(x, y) \equiv \begin{cases} (\frac{1}{y-x} \int_x^y \frac{ds}{m_\sigma(s)})^{-1}, & x \neq y \\ m_\sigma(x), & x = y \end{cases} \tag{C.16}$$

Afterwards, the matrix is transformed back from $\widehat{E}$ to E via $M \equiv A\widehat{M}A^{-1}$. The discretization procedure is then closed by expressing the term

$$M_\sigma \left(U^{k+1} \nabla P^{k+1}, \nabla \Theta \right) \tag{C.17}$$

with respect to the base functions ϕ_i in the form

$$L_l^M(H) \equiv \left(\left(M_\sigma(H) \nabla \phi_i, \nabla \phi_j \right) \right)_{i,j=1}^{\dim V^l} \tag{C.18}$$

so that finally the equation to solve can be written as

$$H^{k+1} - H^k + \tau_k M_l^{-1} L_l^M(H^{k+1}) \left[M_l^{-1} L_l H^{k+1} + \mathcal{T}_l \Phi'_+(H^{k+1}) \right.$$
$$\left. + \mathcal{T}_l \Phi'_-(H^k) \right] = 0, \tag{C.19}$$

with the base scalars

$$M_h \equiv \left((\phi_i, \phi_j)_l \right)_{i,j=1}^{\dim V^l} \tag{C.20}$$

and

$$L_h \equiv \left((\nabla \phi_i, \nabla \phi_j)_l \right)_{i,j=1}^{\dim V^l}. \tag{C.21}$$

The final step then is to set up a *Newton method* to solve the fixed-point problem

$$B(H_{i+1}^{k+1}) = 0 \tag{C.22}$$

for

$$B(H_{i+1}^{k+1}) \equiv H_{i+1}^{k+1} - H^k + \tau_k M_l^{-1} L_l^M(H_i^{k+1}) \left[M_l^{-1} L_l H_{i+1}^{k+1} \right.$$
$$\left. + \mathcal{T}_l \Phi'_+(H_{i+1}^{k+1}) + \mathcal{T}_l \Phi'_-(H^k) \right], \tag{C.23}$$

setting $H_{i+1,0}^{k+1} = H_i^{k+1}$.

Appendix D
Towards 'Better' Theories of Viscoelastic Thin Films

In this final appendix we close the book with a look further on to better models in the hydrodynamic description for non-Newtonian fluids. All our discussion was based on a phenomenological approach by employing the *Jeffreys model*. We have seen that a more 'microscopic' knowledge is required for a comparison with experiment. In this appendix we ask whether we can we make our models more realistic for the problems under study, or at least justify them better.

We go into two aspects of this question, both of which are related to better descriptions of the fluid's stress tensor, as it enters into momentum equation, the Navier-Stokes equation. The first tries to go more fundamental in a macroscopic hydrodynamic description in the sense that it tries to exploit basic (symmetry) principles, while the second employs a kinetic approach in which one tries to build continuum models from more microscopic theories.

D.1 A Nonlinear Fluid Dynamics of Non-Newtonian Fluids

Pleiner et al. (2000, 2004), Temmen et al. (2000) have developed a nonlinear dynamic theory for non-Newtonian fluids based on hydrodynamic principles. If one starts out as usual from the momentum equation, the Navier-Stokes equation in the form

$$\varrho \frac{Dv_i}{Dt} + \nabla_i p + \nabla_j \tau_{ij} = 0 \tag{D.1}$$

where p is the isotropic pressure, one can write the stress tensor τ_{ij} as

$$\tau_{ij} = -\Psi_{ij} + \Psi_{ki} U_{jk} + \Psi_{kj} U_{ik} + \tau_{ij}^{ph} \tag{D.2}$$

where Ψ_{ij} is the *elastic stress tensor*, the thermodynamic conjugate to the strain tensor. Here,

$$U_{ij} = \frac{1}{2} \left[\nabla_j u_i + \nabla_i u_j - (\nabla_i u_k)(\nabla_j u_k) \right] \tag{D.3}$$

R. Blossey, *Thin Liquid Films*, Theoretical and Mathematical Physics, DOI 10.1007/978-94-007-4455-4, © Springer Science+Business Media Dordrecht 2012

is the *Eulerian strain tensor*, where $u_i(r)$ is the displacement field. It fulfills the dynamic equation

$$\frac{D}{Dt}U_{ij} - A_{ij} + U_{ki}\nabla_j v_k + U_{kj}\nabla_i v_k = X_{ij}^{ph} \tag{D.4}$$

with

$$A_{ij} = \frac{1}{2}[\nabla_i v_j + \nabla_j v_i]. \tag{D.5}$$

Further, X_{ij}^{ph} is a phenomenological (quasi-)current containing relaxational processes, e.g., diffusions.

Returning to Eq. (D.2), in this construction, the three terms involving Ψ, both linear and nonlinear, are counterterms to the flows in Eq. (D.4), needed to avoid entropy productions due to reversibility. In this formulation these terms are thus generic; by contrast, the last term is *phenomenological* and reflects material properties. Thus, material properties require to express explicit models for

$$X_{ij}^{ph} = -\alpha_{ijkl}\Psi_{kl} \tag{D.6}$$

and

$$\tau_{ij}^{ph} = -\nu_{ijkl}A_{kl} \tag{D.7}$$

describing strain relaxation and viscosity, respectively, which also admit further generalizations via cross-couplings. Further, one has the relation between strain and elastic stress

$$\Psi_{ij} = K_{ijkl}U_{kl} \tag{D.8}$$

with the elastic tensor K_{ij}.

The tensors $\widehat{\alpha}$ and $\widehat{\nu}$ can also be expanded in the strain tensor $\widehat{U}$, and symmetries exploited for these expressions, a point we do not pursue here any further in all generality. Instead, we briefly show how the expressions can be based on the stress tensor and its derivatives, rather than by strain. This can be done by the corresponding expansions in $\widehat{U}$, but taken only to second order. The dynamic strain equation can then be written as

$$\frac{D}{Dt}U_{ij} - A_{ij} + U_{ki}\nabla v_k + U_{kj}\nabla_i v_k = -\frac{1}{\tau_1}U_{ij} - \frac{1}{\tau_2}U_{ik}U_{jk} \tag{D.9}$$

where $\tau_1^{-1} = \alpha_1 K_1$ and $\tau_2^{-1} = 2\alpha_1 K_2 + 2\alpha_2 K_1$, with elastic moduli K_1, K_2, and the other four parameters deriving from the expansions of the tensors $\widehat{\alpha}$ and $\widehat{\nu}$. The stress tensor τ_{ij} in the Navier-Stokes equation is given by, following Eq. (D.2) and the expansions

$$\tau_{ij} = -K_1 U_{ij} + K_2' U_{ik}U_{jk} - \nu_1 A_{ij} - \nu_2(U_{ik}A_{jk} + U_{jk}A_{ik}) \tag{D.10}$$

with $K_2' = 2(K_1 - K_2)$. Taking the derivative D/Dt of the stress tensor yields

$$\frac{D}{Dt}\tau_{ij} = -F\left[U_{ij}, A_{ij}, \frac{D}{Dt}A_{ij}, \Omega_{ij}\right] \tag{D.11}$$

where Ω_{ij} is the vorticity which we encountered earlier already,

$$\Omega_{ij} = \frac{1}{2}[\nabla_j v_i - \nabla_i v_j]. \tag{D.12}$$

To end up with an equation for τ_{ij}, one needs to invert Eq. (D.10), which can only be done approximately by expanding in U_{ij}, leading to terms linear and quadratic in A_{ij}. This leads to the expressions

$$\tau_1 \frac{D_a}{Dt}\tau_{ij} + \tau_{ij} = -\nu_\infty A_{ij} - \nu_1 \tau_1 \frac{D_b}{Dt} A_{ij} + \frac{r}{K_1}\tau_{ik}\tau_{jk}$$
$$+ \frac{\nu_1 \nu_2}{K_1}\left([\tau_{jk} + \nu_1 A_{jk}]\frac{\partial}{\partial t}A_{ij} + [\tau_{ik} + \nu_1 A_{ik}]\frac{\partial}{\partial t}A_{jk}\right)$$
$$+ \mathcal{O}(3), \tag{D.13}$$

where $\nu_\infty = \nu_1 + \tau_1 K_1$ and $r = \tau_1/\tau_2 - K_2'/K_1$ and the derivative is defined by its action on any tensor T_{ij} to be

$$\frac{D_a}{Dt}T_{ij} \equiv \frac{D}{Dt}T_{ij} - a(T_{ik}A_{jk} + T_{jk}A_{ik}) - (T_{ik}\Omega_{jk} + T_{jk}\Omega_{ik}). \tag{D.14}$$

For $a = -1$ ($a = +1$), D_a/Dt is the lower (upper) convected derivative. If $a = 0$, we have the *Jaumann* or *corotational derivative* which we encountered before. For general a and b one has the material-dependent behaviours

$$a = -1 + \frac{\nu_1}{K_1\tau_2} - \frac{K_2'}{K_1^2}\frac{\nu_\infty}{\tau_1} \tag{D.15}$$

and

$$b = -1 + \frac{\nu_1}{2K_1\tau_2} - \frac{K_2'}{2K_1^2}\frac{\nu_\infty}{2K_1} - \frac{K_2'}{2K_1} - \frac{\nu_2}{\nu_1} \tag{D.16}$$

A final Task. Identify the Maxwell and Jeffreys models from this more general description.

D.2 The Rolie-Poly Equation

In the second approach, one starts out from a microscopic level in order to arrive at macroscopic quantities, like the stress tensor (Byron Bird et al. 1987b). The well-accepted starting point in this context is the *Doi-Edwards tube model*. The basic quantity here is a *Langevin equation* for the motion of a single primitive chain whose motion is confined to a tube. The primitive chain is composed of effective segments, representing sections of a real chain, whereby each tube segment is understood as being the distance between entanglement points, and sufficiently large to have properties of a Gaussian coil.

In this theory, different relaxation mechanism have been built into. The fastest relaxation mode is *chain retraction*, which occurs on the Rouse time, τ_R. It corresponds to a return of the chain to equilibrium under flow deformations of the tube.

Reptation, on the other hand, is the slowest mode, a one-dimensional Brownian motion along the tube, with a timescale τ_d. It dominates in linear chains at very small strain rates, $\dot{\gamma} < \tau_d^{-1}$. Contour length fluctuations of the tube are stochastic, breathing-mode retractions of the chain ends. *Constraint release* is a relaxation mode of the tube due to disappearance of entanglement points when the chain relax out of their tube by reptation of contour length fluctuations. A particular such mode is convective constraint release (CCR), which occurs at shear strain rates larger than the reptation rate, $\dot{\gamma} > \tau_d^{-1}$, when the entanglement are removed with the convective flow of the surrounding chains.

In order to model such effects, and in particular to arrive at a stress tensor description, the stochastic equation of the tube reads as (Likhtman and Graham 2003)

$$\mathbf{R}(s, t + \Delta t) = \mathbf{R}(s + \Delta\xi, t)$$
$$+ \Delta t \left(\kappa_v \cdot \mathbf{R} + \frac{3}{2} \frac{v}{|R'|} \mathbf{R}'' + \mathbf{g}(s, t) + \frac{1}{2\pi^2 \tau_R} \frac{(\mathbf{R}'' \cdot \mathbf{R}')\mathbf{R}'}{R'^2} \right). \tag{D.17}$$

Here, $\mathbf{R}(s, t)$ is a stochastic vector for the position of the tube segment; s labels the monomers in the tube, measured in entanglement segments from one end of the chain, $s = 0, \ldots, Z$, with $Z = N/N_e$. The first term in Eq. (D.17) describes reptation with a noise with variance

$$\langle \Delta\xi(t) \Delta\xi(t') \rangle = 2D_c \delta(t - t'). \tag{D.18}$$

The second term in Eq. (D.17) describes deformation by flow with κ_v as the velocity gradient tensor, the third and forth describe convective constraint release, and the last retraction along the tube contour due to stretch relaxation.

The term $\mathbf{g}$ is stochastic with variance

$$\langle \mathbf{g}(s, t)\mathbf{g}(s', t') \rangle = \frac{\mathbf{I}}{|R'|} v a^2 \delta(s - s')\delta(t - t') \tag{D.19}$$

where $\mathbf{I} = \delta_{ij}$. v is the frequency of constraint release. Further, τ_R is the *Rouse relaxation time* of one entanglement segment, and $D_c = 1/(3\pi^2 Z \tau_R)$ the reptation diffusion constant.

From this equation, a partial differential equation for the tangent correlation function

$$f_{ij}(s, s', t) \equiv \left\langle \frac{\partial R_i}{\partial s} \frac{\partial R_j}{\partial s'} \right\rangle \tag{D.20}$$

can be obtained. It allows to derive the polymeric contribution to the stress tensor

$$\tau_{ij}^{poly} = \frac{3G_e}{Z} \int_0^Z ds\, f_{ij}(s, s). \tag{D.21}$$

The equation can be simplified in order to generate a so-called 'single-mode' equation, which runs under the name *ROuse LInear Entangled POLYmers*—Roly-Poly.

Its form is in an appropriate approximation for the non-stretching case

$$\frac{d}{dt}\tau = \kappa_v^T \cdot \tau + \tau \cdot \kappa_a^T - \frac{1}{\tau_d}(\tau - \mathbf{I}) - \frac{2}{3}tr(\kappa_v^T \cdot \tau)\big(\tau + \beta(\tau - \mathbf{I})\big) \qquad (D.22)$$

where β is a CCR-coefficient. The Rolie-Poly approach has proved successful in the description of macroscopic flows and there are also indications towards the importance of CCR in thin films, see Roth et al. (2005).

Bibliography

Archer, A.J., Robbins, M.J., Thiele, U.: Dynamical density functional theory for the dewetting of evaporating thin films of nanoparticle suspensions exhibiting pattern formation. Phys. Rev. E **81**, 021602 (2010)

Atena, A., Khenner, M.: Thermocapillary effects in driven dewetting and self assembly of pulsed-laser-irradiated metallic films. Phys. Rev. B **80**, 075402 (2009)

Ausserré, D., Brochard-Wyart, F., de Gennes, P.G.: Dewetting of an incompressible, stratified fluid. C. R. Acad. Sci. Paris **320**, 131–136 (1995)

Bäumchen, O., Fetzer, R., Jacobs, K.: Reduced interfacial entanglement density affects the boundary conditions of polymer flow. Phys. Rev. Lett. **103**, 247801 (2009)

Bäumchen, O., Jacobs, K.: Slip effects in polymer thin films. J. Phys., Condens. Matter **22**, 033102 (2010)

Bauer, C., Dietrich, S.: Quantitative study of laterally inhomogeneous wetting films. Eur. Phys. J. B **10**, 767–779 (1999)

Bausch, R., Blossey, R.: Critical droplets in first-order wetting transitions. Europhys. Lett. **14**, 125–129 (1991)

Bausch, R., Blossey, R.: Critical droplets at a wall near a first-order wetting transition. Phys. Rev. E **48**, 1131–1135 (1993)

Bausch, R., Blossey, R., Burschka, M.A.: Critical nuclei for wetting and dewetting. J. Phys. A **27**, A1405–A1406 (1994)

Bausch, R., Blossey, R.: Lifetime of undercooled wetting layers. Phys. Rev. E **50**, R1759–R1761 (1994)

Becker, J., Grün, G., Seemann, R., Mantz, H., Jacobs, K., Mecke, K.R., Blossey, R.: Complex dewetting scenarios captured by thin-film models. Nat. Mater. **2**, 59–63 (2003)

Becker, J., Grün, G.: The thin-film equation: recent advances and some new perspectives. J. Phys., Condens. Matter **17**, S291–S307 (2005)

Berestycki, H., Lions, P.L., Peletier, L.A.: An O.D.E. approach to the existence of positive solutions for semi-linear problems. Indiana Univ. Math. J. **30**, 141–157 (1981)

Bergeron, V., Jiménez-Laguna, A.I., Radke, C.J.: Hole formation and sheeting in the drainage of thin liquid films. Langmuir **8**, 3027–3032 (1992)

Bischof, J., Scherer, D., Herminghaus, S., Leiderer, P.: Dewetting modes of thin metallic films: nucleation of holes and spinodal dewetting. Phys. Rev. Lett. **77**, 1536–1539 (1996)

Blossey, R.: Nucleation at first-order wetting transitions. Int. J. Mod. Phys. B **9**, 3489–3525 (1995)

Blossey, R., Indekeu, J.O.: Interface-potential approach to surface states in type-I superconductors. Phys. Rev. B **53**, 8599–8603 (1996)

Blossey, R.: Dimple-assisted dewetting in rotating superfluid films. Phys. Rev. B **57**, R14048–R14051 (1998)

R. Blossey, *Thin Liquid Films*, Theoretical and Mathematical Physics,
DOI 10.1007/978-94-007-4455-4, © Springer Science+Business Media Dordrecht 2012

Blossey, R., Oligschleger, C.: First-order wetting transitions under gravity. J. Colloid Interface Sci. **209**, 442–444 (1999)

Blossey, R.: Dimple-assisted dewetting: heterogeneous nucleation in undercooled wetting films. Ann. Phys. (Leipz.) **10**, 733–775 (2001a)

Blossey, R.: Effective forces between interfaces in type-I superconductors. Europhys. Lett. **54**, 522–525 (2001b)

Blossey, R., Münch, A., Rauscher, M., Wagner, B.: Slip vs. viscoelasticity in dewetting thin films. Eur. Phys. J. E **20**, 267–272 (2006)

Blossey, R.: Thin film rupture and polymer flow. Phys. Chem. Chem. Phys. **10**, 5177–5183 (2008)

Böhme, G.: Strömungsmechanik nichtnewtonscher Fluide. Teubner Verlag, Stuttgart (2000)

Bonn, D., Kellay, H., Wegdam, G.H.: Experimental observation of hysteresis in a wetting transition. Phys. Rev. Lett. **69**, 1975–1978 (1992)

Bonn, D., Eggers, J., Indekeu, J.O., Meunier, J., Rolley, E.: Wetting and spreading. Rev. Mod. Phys. **81**, 739–805 (2009)

Brinkmann, M., Blossey, R.: Blobs, channels and "cigars": morphologies of liquids at a step. Eur. Phys. J. E **14**, 79–89 (2004)

Brochard, F., Redon, C., Rondelez, F.: Dewetting: the gravity controlled regime. C. R. Acad. Sci. Paris **306**, 1143–1146 (1988)

Brochard-Wyart, F., Daillant, J.: Drying of solids wetted by thin liquid films. Can. J. Phys. **68**, 1084–1088 (1989)

Brochard-Wyart, F., di Meglio, J.-M., Quéré, D., de Gennes, P.G.: Spreading of nonvolatile liquids in a continuum picture. Langmuir **91**, 335–338 (1991)

Brochard-Wyart, F., Redon, C.: Dynamics of liquid rim instabilities. Langmuir **8**, 2324–2329 (1992)

Brochard, F., de Gennes, P.-G.: Shear-dependent slippage at a polymer/solid interface. Langmuir **8**, 3033–3037 (1992)

Brochard-Wyart, F., Martin, P., Redon, C.: Liquid/liquid dewetting. Langmuir **9**, 3682–3690 (1993)

Brochard-Wyart, F., de Gennes, P.G., Hervert, H., Redon, C.: Wetting and slippage of polymer melts on semi-ideal surfaces. Langmuir **10**, 1566–1572 (1994a)

Brochard-Wyart, F., de Gennes, P.G.: Dewetting of a water film between a solid and a rubber. J. Phys., Condens. Matter **6**, A9–A12 (1994b)

Brochard-Wyart, F., Buguin, A.: Shape effects in inertial dewetting. C. R. Acad. Sci. Paris **327**, 809–815 (1999)

Brzoska, J.B., Benazouz, I., Rondelez, F.: Silanization of solid substrates: a step toward reproducibility. Langmuir **10**, 4367–4373 (1994)

Burelbach, J.P., Bankoff, S.G., Davis, S.H.: Nonlinear stability of evaporating/condensing films. J. Fluid Mech. **195**, 463–494 (1988)

Byron Bird, R., Armstrong, R.C., Hassager, O.: Dynamics of Polymeric Liquids, Vol. 1: Fluid Mechanics, 2nd edn. Wiley, New York (1987a)

Byron Bird, R., Curtiss, C.F., Armstrong, R.C., Hassager, O.: Dynamics of Polymeric Liquids, Vol. 2: Kinetic Theory, 2nd edn. Wiley, New York (1987b)

Cahn, J.W.: Critical point wetting. J. Chem. Phys. **66**, 3667–3672 (1977)

Casteletto, V., Cantat, I., Sarker, D., Bausch, R., Bonn, D., Meunier, J.: Stability of soap films: hysteresis and nucleation of black films. Phys. Rev. Lett. **90**, 048302 (2003)

Cheng, E., Cole, M.W., Dupont-Roc, J., Saam, W.F., Treiner, J.: Novel wetting behavior in quantum films. Rev. Mod. Phys. **65**, 557–567 (1993)

Coleman, S.: In: Aspects of Symmetry: The Uses of Instantons. Cambridge University Press, Cambridge (1985)

Craster, R.V., Matar, O.K.: Dynamics and stability of thin liquid films. Rev. Mod. Phys. **81**, 1131–1198 (2009)

Dal Passo, R., Garcke, H., Grün, G.: On a fourth-order degenerate parabolic equation: global entropy estimates and qualitative behaviour of solutions. SIAM J. Math. Anal. **29**, 321–342 (1998)

Damman, P., Gabriele, S., Coppée, S., Descprez, S., Villers, D., Vilmin, T., Raphaël, E., Hamieh, M., Al Akhrass, S., Reiter, G.: Relaxation of residual stress and reentanglement of polymers in spin-coated films. Phys. Rev. Lett. **99**, 036101 (2007)

Deegan, R.D., Bakajin, O., Dupont, T.F., Huber, G., Nagel, S.R., Witten, T.A.: Capillary flow as the cause of ring stains from dried liquid drops. Nature **389**, 827–829 (1997)

Deegan, R.D., Bakajin, O., Dupont, T.F., Huber, G., Nagel, S.R., Witten, T.A.: Contact line deposits in an evaporating drop. Phys. Rev. E **62**, 756–765 (2000)

De Feijter, J.A., Vrij, A.: Transition regions. Line tensions and contact angles in soap films. J. Electroanal. Chem. **37**, 9–22 (1972)

de Gennes, P.G.: Wetting: statics and dynamics. Rev. Mod. Phys. **57**, 827–863 (1985)

de Gennes, P.G.: Dewetting between a porous solid and a rubber. C. R. Acad. Sci. Paris **318**, 1033–1037 (1994)

de Gennes, P.G.: Dry vortices in thin helium films. C. R. Acad. Sci. Ser. 2 **327**, 1337–1343 (1999)

de Gennes, P.G.: Glass transition in thin polymer films. Eur. Phys. J. E **2**, 201–205 (2000)

de Gennes, P.G., Brochard-Wyart, F., Quéré, D.: Gouttes, Bulles, Perles et Ondes. Belin, Paris (2005)

Derjaguin, B.V.: On the thickness of the liquid film adhering to the walls of a vessel after emptying. Acta Physicochim. USSR **20**, 349–352 (1943)

de Ryck, A., Quéré, D.: Inertial coating of a fibre. J. Fluid Mech. **311**, 219–237 (1996)

Dietzel, M., Troian, S.M.: Formation of nanopillar arrays in ultrathin viscous films: the critical role of thermocapillary stresses. Phys. Rev. Lett. **103**, 074501 (2009)

Dobbs, H.T., Indekeu, J.O.: Line tension at wetting: interface displacement model beyond the squared-gradient approximation. Physica A **201**, 457–481 (1993)

Dobbs, H.: The modified Young's equation for the contact angle of a small sessile drop from an interface displacement model. Int. J. Mod. Phys. B **13**, 3255–3259 (1999a)

Dobbs, H.: The elasticity of a contact line. Physics A **271**, 36–47 (1999b)

Dobbs, H., Blossey, R.: Capillary-wave effects at critical wetting in type-I superconductors. Phys. Rev. E **61**, R6049–R6051 (2000)

Dzyaloshinskii, I.E., Lifshitz, E.M., Pitaevskii, L.P.: The general theory of van der Waals forces. Adv. Phys. **10**, 165–209 (1961)

Eggers, J.: Dynamics of liquid nanojets. Phys. Rev. Lett. **89**, 084502 (2002)

Evans, R.: The nature of the liquid-vapour interface and other topics in the statistical mechanics of non-uniform, classical fluids. Adv. Phys. **28**, 143–200 (1979)

Fakhraai, Z., Forrest, J.A.: Measuring the surface dynamics of glassy polymers. Science **319**, 600–604 (2008)

Fetzer, R., Jacobs, K., Münch, A., Wagner, B., Witelski, T.P.: New slip regimes and the shape of dewetting thin liquid films. Phys. Rev. Lett. **95**, 127801 (2005)

Fetzer, R., Rauscher, M., Münch, A., Wagner, B.A., Jacobs, K.: Slip-controlled thin-film dynamics. Europhys. Lett. **75**, 638–644 (2006)

Fetzer, R., Rauscher, M., Seemann, R., Jacobs, K., Mecke, K.: Thermal noise influences fluid flow in thin films during spinodal dewetting. Phys. Rev. Lett. **99**, 114503 (2007a)

Fetzer, R., Münch, A., Wagner, B., Rauscher, M., Jacobs, K.: Quantifying hydrodynamic slip: a comprehensive analysis of dewetting profiles. Langmuir **23**, 10559–10566 (2007b)

Flitton, J., King, J.R.: Surface-tension driven dewetting of Newtonian and power-law fluids. J. Eng. Math. **50**, 241–266 (2004)

Foltin, G., Bausch, R., Blossey, R.: Critical holes in undercooled wetting layers. J. Phys. A **30**, 2937–2946 (1997)

Fowlkes, J.D., Wu, Y., Rack, P.D.: Directed assembly of bimetallic nanoparticles by pulsed-laser-induced dewetting: a unique time and length scale regime. Appl. Mater. & Interfaces **2**, 2153–2161 (2010)

Fradin, C., Braslau, A., Luzet, D., Smilgies, D., Alba, M., Boudet, N., Mecke, K., Daillant, J.: Reduction in the surface energy of liquid interfaces at short length scales. Nature **403**, 871–874 (2000)

Gabriele, S., Damman, P., Sclavons, S., Desprez, S., Coppée, S., Reiter, G., Hamieh, M.,
    Al Akhrass, S., Vilmin, T., Raphäel, E.: Viscoelastic dewetting of constrained polymer thin
    films. J. Polym. Sci. **44**, 3022–3030 (2006a)
Gabriele, S., Sclavons, S., Reiter, G., Damman, P.: Disentanglement time of polymers determines
    the onset of rim instabilities in dewetting. Phys. Rev. Lett. **96**, 156105 (2006b)
Grün, G., Rumpf, M.: Nonnegativity-preserving convergent schemes for the thin-film equation.
    Numer. Math. **87**, 113–152 (2000)
Grün, G., Mecke, K., Rauscher, M.: Thin-film flow influenced by thermal noise. J. Stat. Phys. **122**,
    1261–1291 (2006)
Gutfreund, P., Bäumchen, O., van der Grinten, D., Fetzer, R., Maccarini, M., Jacobs, K., Zabel,
    H., Wolff, M.: Surface correlation affects liquid order and slip in a Newtonian liquid (2011).
    arXiv:1104.0868
Hamieh, M., Al Akhrass, S., Hamieh, T., Damman, P., Gabriele, S., Vilmin, T., Raphaël, E., Reiter,
    G.: Influence of substrate properties on the dewetting dynamics of viscoelastic polymer films.
    J. Adhes. **83**, 367–381 (2007)
Herminghaus, S., Jacobs, K., Mecke, K., Bischof, J., Fery, A., Ibn-Elhaj, M., Schlagowski, S.:
    Spinodal dewetting in liquid crystal and liquid metal films. Science **282**, 916–919 (1998)
Herminghaus, S., Fery, A., Schlagowski, S., Jacobs, K., Seemann, R., Gau, H., Mönch, W.,
    Pompe, T.: Liquid microstructures at solid surfaces. J. Phys., Condens. Matter **11**, A57–A74
    (1999)
Herminghaus, S.: Polymer thin films and surfaces: possible effects of capillary waves. Eur. Phys.
    J. E **8**, 237–243 (2002)
Herminghaus, S., Jacobs, K., Seemann, R.: Viscoelastic dynamics of polymer thin films. Eur. Phys.
    J. E **12**, 101–110 (2003)
Herminghaus, S., Brochard, F.: Dewetting through nucleation. C. R. Phys. **7**, 1073–1081 (2006).
    Correction: C. R. Phys. **8**, 86 (2007)
Indekeu, J.O., van Leeuwen, J.M.J.: Interface delocalization transition in type-I superconductors.
    Phys. Rev. Lett. **75**, 1618–1621 (1995)
Israelachvili, J.: Interfacial and Surface Forces, 2nd edn. Academic Press, London (1992)
Jacobs, K., Herminghaus, S., Mecke, K.R.: Thin liquid polymer films rupture via defects. Langmuir
    14, 965–969 (1998a)
Jacobs, K., Seemann, R., Schatz, G., Herminghaus, S.: Growth of holes in liquid films with partial
    slippage. Langmuir **14**, 4961–4963 (1998b)
Joanny, J.F., de Gennes, P.G.: Nucleation under conditions of complete wetting. C. R. Acad. Sci.
    303, 337–340 (1984)
Joanny, J.-F., Prost, J.: Active gels as a description of the actin-myosin cytoskeleton. HFSP J. **3**,
    94–104 (2009)
Kargupta, K., Konnur, R., Sharma, A.: Instability and pattern formation in thin liquid films on
    chemically heterogeneous substrates. Langmuir **16**, 10243–10253 (2000)
Kargupta, K., Sharma, A.: Dewetting of thin films on periodic physically and chemically patterned
    surfaces. Langmuir **18**, 1893–1903 (2002)
Kargupta, K., Sharma, A., Khanna, R.: Instability, dynamics and morphology of slipping thin films.
    Langmuir **20**, 244–253 (2004)
Kellay, H., Bonn, D., Meunier, J.: Prewetting in a binary liquid mixture. Phys. Rev. Lett. **71**, 2607–
    2610 (1993)
Kerle, T., Klein, J., Yerushalmi-Rozen, R.: Accelerated rupture at the liquid/liquid interface. Lang-
    muir **18**, 10146–10154 (2002)
Khayat, R.E.: Transient two-dimensional coating flow of a viscoelastic fluid film on a substrate of
    arbitrary shape. J. Non-Newton. Fluid Mech. **95**, 199–233 (2001)
King, J.R., Münch, A., Wagner, B.: Linear stability of a ridge. Nonlinearity **19**, 2813–2831 (2006)
Kondic, L., Diez, J.A., Rack, P.D., Guan, Y., Fowlkes, J.D.: Nanoparticle assembly via the dewet-
    ting of patterned thin metal lines: understanding the instability mechanism. Phys. Rev. E **79**,
    026302 (2009)

Kozhevnikov, V.F., van Bael, M.J., Sahoo, P.K., Temst, K., van Haesendonck, C., Vantomme, A., Indekeu, J.O.: Observation of wetting-like phase transitions in a surface-enhanced type-I superconductor. New J. Phys. **9**, 75 (2007)

Kruse, K., Joanny, J.F., Jülicher, F., Prost, J., Sekimoto, K.: Generic theory of active polar gels: a paradigm for cytoskeletal dynamics. Eur. Phys. J. **16**, 5–16 (2005)

Landau, L.D., Levich, B.V.: Dragging of a liquid by a moving plate. Acta Physicochim. USSR **17**, 42–54 (1942)

Landau, L.D., Lifshitz, E.M.: Fluid Dynamics. Butterworth/Heinemann, Stoneham/London (1987)

Latterini, L., Blossey, R., Hofkens, J., Vanoppen, P., de Schryver, F.C., Rowan, A.E., Nolte, R.J.M.: Ring formation in evaporating porphyrin derivative solutions. Langmuir **15**, 3582–3588 (1999)

Ledesma-Aguilar, R., Hernández-Machado, A., Pagonabarraga, I.: Dynamics of gravity driven three-dimensional thin films on hydrophilic and hydrophobic patterned substrates. Langmuir **26**, 3292–3301 (2009)

Léger, L., Hervet, H., Bureau, L.: Friction mechanisms at polymer-solid interfaces. C. R., Chim. **9**, 80–89 (2006)

Likhtman, A.E., Graham, R.S.: Simple constitutive equation for linear polymer melts derived from molecular theory: Rolie-Poly equation. J. Non-Newton. Fluid Mech. **114**, 1–12 (2003)

Lin, Z., Kerle, T., Russell, T.P., Schäffer, E., Steiner, U.: Electric field induced dewetting at polymer/polymer interfaces. Macromolecules **35**, 6255–6262 (2002)

Martin, A., Buguin, A., Brochard-Wyart, F.: "Cerenkov" dewetting at soft interfaces. Europhys. Lett. **57**, 604–610 (2002)

Martin, P., Brochard-Wyart, F.: Dewetting at soft interfaces. Phys. Rev. Lett. **80**, 3296–3299 (1998)

Mecke, K.R., Dietrich, S.: Effective Hamiltonian for liquid-vapor interfaces. Phys. Rev. E **59**, 6766–6784 (1999)

Mecke, K.R., Stoyan, D. (eds.): Statistical Physics and Spatial Statistics—The Art of Analysing and Modelling Spatial Structures and Pattern Formation. Lecture Notes in Physics, vol. 554. Springer, Berlin (2000)

Mecke, K., Rauscher, M.: On thermal fluctuations in thin film flow. J. Phys., Condens. Matter **17**, S3515–S3522 (2005)

Migler, K.B., Hervet, H., Leger, L.: Slip transition of a polymer melt under shear stress. Phys. Rev. Lett. **70**, 287–290 (1993)

Montevecchi, E., Blossey, R.: Heterogeneous hole nucleation in electron-charged helium films. Phys. Rev. Lett. **85**, 4743–4746 (2000)

Morey, F.C.: Thickness of a film adhering to a surface slowly withdrawn from the liquid. J. Res. Natl. Bur. Stand. **25**, 385–393 (1940)

Moseler, M., Landmann, U.: Formation, stability, and breakup of nanojets. Science **289**, 1165–1169 (2000)

Münch, A., Wagner, B., Witelski, T.P.: Lubrication models with small to large slip lengths. J. Eng. Math. **53**, 359–383 (2005)

Münch, A., Wagner, B.: Contact-line instability of dewetting thin films. Physica D **209**, 178–190 (2005)

Münch, A., Wagner, B., Rauscher, M., Blossey, R.: A thin-film model for corotational Jeffreys fluids under strong slip. Eur. Phys. J. E **20**, 365–368 (2006)

Münch, A., Wagner, B.: Impact of slippage on the morphology and stability of a dewetting rim. J. Phys., Condens. Matter **23**, 184101 (2011)

Nacher, P.J., Dupont-Roc, J.: Experimental evidence for nonwetting with superfluid helium. Phys. Rev. Lett. **67**, 2966–2969 (1992)

Nakanishi, H., Fischer, M.E.: Multicriticality of wetting, prewetting, and surface transitions. Phys. Rev. Lett. **49**, 1565–1568 (1982)

Neto, C., Jacobs, K., Seemann, R., Blossey, R., Becker, J., Grün, G.: Satellite hole formation during dewetting: experiment and simulation. J. Phys., Condens. Matter **15**, 3355–3366 (2003)

Ohara, P.C., Heath, J.R., Gelbart, W.M.: Bildung von Submikrometer-grossen Partikelringen bem Verdunsten Nanopartikelhaltiger Lösungen. Angew. Chem. **109**, 1120–1122 (1997)

Ohara, P.C., Gelbart, W.: Interplay between hole instability and nanoparticle array formation in ultrathin liquid films. Langmuir **14**, 3418–3424 (1997)

Oron, A., Davis, S.H., Bankoff, S.G.: Long-scale evolution of thin liquid films. Rev. Mod. Phys. **69**, 931–980 (1997)

Parisse, F., Allain, C.: Shape changes of colloidal droplets during drying. J. Phys. II France **6**, 1111–1119 (1996)

Pleiner, H., Liu, M., Brand, H.R.: The structure of convective nonlinearities in polymer rheology. Rheol. Acta **39**, 560–565 (2000)

Pleiner, H., Liu, M., Brand, H.R.: The structure of convective nonlinearities in polymer rheology. Rheol. Acta **43**, 502–508 (2004)

Pompe, T., Herminghaus, S.: Three-phase contact line energetics from nanoscale liquid surface topographies. Phys. Rev. Lett. **85**, 1930–1933 (2000)

Rauscher, M., Münch, A., Wagner, B., Blossey, R.: A thin-film equation for viscoelastic liquids of Jeffreys type. Eur. Phys. J. E **17**, 373–379 (2005)

Rauscher, M., Blossey, R., Münch, A., Wagner, B.: Spinodal dewetting of thin films with large interfacial slip: implications from the dispersion relation. Langmuir **24**, 12290–12294 (2008)

Redon, C., Brochard-Wyart, F., Rondelez, F.: Dynamics of dewetting. Phys. Rev. Lett. **66**, 715–718 (1991)

Reiter, G.: Dewetting of thin polymer films. Phys. Rev. Lett. **68**, 75–78 (1992)

Reiter, G.: Dewetting of highly elastic thin polymer films. Phys. Rev. Lett. **87**, 186101 (2001)

Reiter, G., Hamieh, M., Damman, P., Sclavons, S., Gabriele, S., Vilmin, T., Raphaël, E.: Residual stresses in thin polymer films cause rupture and dominate early stages of dewetting. Nat. Mater. **4**, 754–758 (2005)

Reiter, G., Al Akhrass, S., Hamieh, M., Damman, P., Gabriele, S., Vilmin, T., Raphaël, E.: Dewetting as an investigative tool for studying properties of thin polymer films. Eur. Phys. J. Spec. Top. **166**, 165–172 (2009)

Ross, D., Bonn, D., Meunier, J.: Observation of short-range critical wetting. Nature **400**, 737–739 (1999)

Roth, C.B., Deh, B., Nickel, B.G., Dutcher, J.R.: Evidence of convective constraint release during hole growth in freely standing polystyrene films at low temperatures. Phys. Rev. E **72**, 021802 (2005)

Roth, C.B., Dutcher, J.R.: Glass transition and chain mobility in thin polymer films. J. Electroanal. Chem. **584**, 13–22 (2005)

Rutledge, J.E., Taborek, P.: Novel wetting behaviour of ^{4}He on Cs. Phys. Rev. Lett. **68**, 2184–2187 (1992)

Samid-Merzel, N., Lipson, S.G., Tannhauser, D.S.: Pattern formation in drying water films. Phys. Rev. E **57**, 2906–2913 (1998)

Schäffer, E., Thurn-Albrecht, T., Russell, T.P.: Electrically induced structure formation and pattern transfer. Nature **403**, 874–877 (2000)

Schäffer, E., Thurn-Albrecht, T., Russell, T.P., Steiner, U.: Electrohydrodynamic instabilities in polymer films. Europhys. Lett. **53**, 518–524 (2001)

Schäffer, E., Harkema, S., Roerdink, M., Blossey, R., Steiner, U.: Morphological instability of a confined polymer film in a thermal gradient. Macromolecules **36**, 1645–1655 (2003)

Schenning, A.P.H.J., Benneker, F.B.G., Geurts, H.P.M., Liu, X.Y., Nolte, R.J.M.: Porphyrin wheels. J. Am. Chem. Soc. **118**, 8549–8552 (1996)

Schick, M., Taborek, P.: Anomalous nucleation at first-order wetting transitions. Phys. Rev. B **46**, 7312–7314 (1992)

Seemann, R., Jacobs, K., Blossey, R.: Polystyrene nanodroplets. J. Phys., Condens. Matter **13**, 4915–4923 (2001a)

Seemann, R., Herminghaus, S., Jacobs, K.: Dewetting patterns and molecular forces: a reconciliation. Phys. Rev. Lett. **86**, 5534–5537 (2001b)

Seemann, R., Herminghaus, S., Jacobs, K.: Gaining control of pattern formation of dewetting liquid films. J. Phys., Condens. Matter **13**, 4925–4938 (2001c)

Seemann, R., Herminghaus, S., Jacobs, K.: Shape of a liquid front upon dewetting. Phys. Rev. Lett. **87**, 196101 (2001d). Erratum PRL **87**, 249902

Seemann, R., Herminghaus, S., Neto, C., Schlagowski, S., Podzimek, D., Konrad, R., Mantz, H., Jacobs, K.: Dynamics and structure formation in thin polymer melt films. J. Phys., Condens. Matter **17**, S267–S290 (2005)

Sekimoto, K., Oguma, R., Kawasaki, K.: Morphological stability analysis of partial wetting. Ann. Phys. **176**, 359–392 (1987)

Snoeijer, J.H., Ziegler, J., Andreotti, B., Fermigier, M., Eggers, J.: Thick films of viscous fluid coating a plate withdrawn from a liquid reservoir. Phys. Rev. Lett. **100**, 244502 (2008)

Snoeijer, J.H., Eggers, J.: Asymptotics of the dewetting rim. Phys. Rev. E **82**, 056314 (2010)

Taborek, P., Rutledge, J.E.: Prewetting phase diagram of ^{4}He on Cs. Phys. Rev. Lett. **69**, 937–940 (1992)

Thiele, U., Mertig, M., Pompe, W.: Dewetting of an evaporating thin liquid film: heterogeneous nucleation and surface instability. Phys. Rev. Lett. **80**, 2869–2872 (1998)

Temmen, H., Pleiner, H., Liu, M., Brand, H.R.: Convective nonlinearity in non-Newtonian fluids. Phys. Rev. Lett. **84**, 3228–3231 (2000)

Trice, J., Thomas, D., Favazza, C., Sureshkumar, R., Kalyanaraman, R.: Pulsed-laser-induced dewetting in nanoscopic metal films: theory and experiment. Phys. Rev. E **75**, 235439 (2007)

Trice, J., Favazza, C., Thomas, D., Garcia, H., Kalyanaraman, R., Sureshkumar, R.: Novel self-organization mechanism in ultrathin liquid films: theory and experiment. Phys. Rev. Lett. **101**, 017802 (2008)

Vilmin, T., Raphaël, E.: Dewetting of thin viscoelastic polymer films on slippery substrates. Europhys. Lett. **72**, 781–787 (2005)

Vilmin, T., Raphaël, E.: Dewetting of thin polymer films. Eur. Phys. J. E **21**, 161–174 (2006)

Vilmin, T., Raphaël, E., Damman, P., Sclavons, S., Gabriele, S., Hamieh, M., Reiter, G.: The role of nonlinear friction in the dewetting of thin polymer films. Europhys. Lett. **73**, 906–912 (2006)

Vrij, A.: Possible mechanism for the spontaneous rupture of thin, free liquid films. Discuss. Faraday Soc. **42**, 23–33 (1966)

Weijs, J.H., Marchand, A., Andreotti, B., Lohse, D., Snoeijer, J.H.: Origin of line tension for a Lennard-Jones nanodroplet. Phys. Fluids **23**, 022001 (2011)

Wu, Y., Fowlkes, J.D., Rack, P.D., Diez, J.A., Kondic, L.: On the breakup of patterned nanoscale copper rings into droplets via Pulsed-Laser-Induced Dewetting: competing liquid-phase instability and transport mechanisms. Langmuir **26**, 11972–11979 (2010)

Yang, Z., Fujii, Y., Lee, F.K., Lam, C.-H., Tsui, O.K.C.: Glass transition dynamics and surface layer mobility in unentangled polystyrene films. Science **328**, 1676–1679 (2010)

Yerushalmi-Rozen, R., Kerle, T., Klein, J.: Alternative dewetting pathways of thin liquid films. Science **285**, 1254–1256 (1999)

Zhang, Y.L., Matar, O.K., Craster, R.V.: Surfactant spreading on a thin weakly viscoelastic film. J. Non-Newton. Fluid Mech. **105**, 53–78 (2002)

Ziebert, F., Raphaël, E.: Dewetting of thin polymer films: influence of interface evolution. Europhys. Lett. **86**, 46001 (2009a)

Ziebert, F., Raphaël, E.: Dewetting dynamics of stressed viscoelastic thin polymer films. Phys. Rev. E **79**, 031605 (2009b)

Index

R. Blossey, *Thin Liquid Films*, Theoretical and Mathematical Physics,
DOI 10.1007/978-94-007-4455-4, © Springer Science+Business Media Dordrecht 2012